6th International Conference on Manufacturing Engineering and Process (ICMEP 2017)

Journal of Physics: Conference Series Volume 885

Lisbon, Portugal
27 – 29 May 2017

ISBN: 978-1-5108-4788-0
ISSN: 1742-6588

Printed from e-media with permission by:

Curran Associates, Inc.
57 Morehouse Lane
Red Hook, NY 12571

Some format issues inherent in the e-media version may also appear in this print version.

Printed by Curran Associates, Inc. (2017)

For permission requests, please contact the Institute of Physics at the address below.

Institute of Physics
Dirac House, Temple Back
Bristol BS1 6BE UK

Phone: 44 1 17 929 7481
Fax: 44 1 17 920 0979

techtracking@iop.org

Additional copies of this publication are available from:

Curran Associates, Inc.
57 Morehouse Lane
Red Hook, NY 12571 USA
Phone: 845-758-0400
Fax: 845-758-2633
Email: curran@proceedings.com
Web: www.proceedings.com

Table of contents

Volume 885

The 6th International Conference on Manufacturing Engineering and Process
27–29 May 2017, Lisbon, Portugal

Accepted papers received: 14 August 2017
Published online: 29 August 2017

Preface

011001
OPEN ACCESS
The 6th International Conference on Manufacturing Engineering and Process

011002
OPEN ACCESS
Conference Committee

011003
OPEN ACCESS
Peer review statement

Biological materials and processing technology

012001
OPEN ACCESS
Optimization of factors to obtain cassava starch films with improved mechanical properties

Mayra Monteiro, Victor Oliveira, Francisco Santos, Eduardo Barros Neto, Karyn Silva, Rayane Silva, João Henrique and Abimaelle Chibério.....1

012002
OPEN ACCESS
Statistical analysis of the factors that influenced the mechanical properties improvement of cassava starch films

Mayra Monteiro, Victor Oliveira, Francisco Santos, Eduardo Barros Neto, Karyn Silva, Rayane Silva, João Henrique and Abimaelle Chibério.....6

012003
OPEN ACCESS
Failure analysis of various monolithic posterior aesthetic dental crowns using finite element method

Liliana Porojan and Florin Topală.....11

012004
OPEN ACCESS
Surface texture and hardness of dental alloys processed by alternative technologies

Liliana Porojan, Cristina E. Savencu, Florin I. Topală and Sorin D. Porojan.....16

012005
OPEN ACCESS
Structural and morphological approach of Co-Cr dental alloys processed by alternative manufacturing technologies

Sorin Porojan, Mihaela Bîrdeanu, Cristina Savencu and Liliana Porojan.....21

Analysis and application of material properties

012006
OPEN ACCESS
Kinetic of carbonic anhydrase immobilized onto amberlite xad 7 and it application in sequestration of co_2 into $caco_3$ precipitate

Siti Noor Hazirah Mohd Rafidi, Siti Nadia Abdullah and Fazlena Hamzah.....26

012007
OPEN ACCESS
Determining the size of nanoparticles in the example of magnetic iron oxide core-shell systems

Maciej Jarzębski, Mikołaj Kościński and Tomasz Białopiotrowicz.....31

012008
OPEN ACCESS
Effect of concentration of Curcuma longa L. on chitosan–starch based edible coating

N. M. Yusof, J. Jai, F. Hamzah, A. Yahya and S. Pinijsuwan.....36

012009
OPEN ACCESS
Online particle size distribution estimation of a mixture of similar sized particles with acoustic emissions

Ejay Nsugbe, Andrew Starr, Ian Jennions and Cristobal Ruiz Carcel.....41

Mechanical engineering and computer aided design

012010
OPEN ACCESS
Multi objective Taguchi optimization approach for resistance spot welding of cold rolled TWIP steel sheets

Mumin Tutar, Hakan Aydin and Ali Bayram.....46

012011
OPEN ACCESS
Investigation on heat transfer characteristic and optimization of the cooling air inlet for the twin-web turbine disk

Mengchuang Zhang, Wenxuan Gou, Qin Yao, Lei Li and Zhufeng Yue.....51

Manufacturing systems and production management

012012
OPEN ACCESS
Effect of deposition speed on the flatness and cylindricity of parts produced by three dimensional printing process

Muhammad Fahad, Mahmood Khalid, Muhammad Nauman and Maqsood Ahmed Khan.....58

012013
OPEN ACCESS
A methodology to determine the level of automation to improve the production process and reduce the ergonomics index

Alejandro Chan-Amaya, María Elena Anaya-Pérez and Víctor Hugo Benítez-Baltazar.....63

012014
OPEN ACCESS
Reduction of defective products by empiric analysis: the production line case in a factory of automotive parts.

Ramses Lugo-Telles and María de los Ángeles Navarrete.....69

012015
OPEN ACCESS
Synchronization of workshops, using facilities planning

Britel Zineb and Cherkaoui Abdelghani.....74

012016
OPEN ACCESS
Simulation based optimized beam velocity in additive manufacturing

Frédéric Vignat, Nicolas Béraud and François Villeneuve.....80

012017
OPEN ACCESS
Drying effects of vacuum far-infrared on aquilaria malaccensis leaves

Habsah Alwi, Sherif Abdulbari Ali, Ku Halim Ku Hamid, Mohamad Zaid Shamsudin, Nurhaslina Che Radzi, Nurul Asyikin Md Zaki and Miradatul Najwa Muhd.....88

Experiment and analysis of mechanical properties of functional materials

012018
OPEN ACCESS
A numerical solution for the axisymmetric flow of a plastic tube on a rigid fibre

Thanh Nguyen, Kien Nguyen and Sergei Alexandrov.....93

012019
OPEN ACCESS
The research on improving the anti-oxidation of tungsten rhenium alloy wires

Xinwei Bo, Qi Liu, Hao Yang, Chundong Jiang, Haoran He, Xiaoyu Wang, Chengchao Liu and Demao Chen.....101

012020
OPEN ACCESS
Laminar heat transfer with alumina nanofluid in cu tube with different diameters

Sedong Kim, A K M Mahmudul Haque, Junhyo Kim, Kitae Yu, Heegeun Song, Cheol Park, Hanshik Chung and Hyomin Jeong.....106

012021
OPEN ACCESS
CFD Analysis of nanofluid forced convection heat transport in laminar flow through a compact pipe

Kitae Yu, Cheol Park, Sedon Kim, Heegun Song and Hyomin Jeong.....112

012022
OPEN ACCESS
Application of $CaCO_3$, CaF_2, SiO_2, and TiO_2 particles to silicone lens for enhancing angular color uniformity of white LED lamp

Anh Q. D. Nguyen and Vinh H. Nguyen.....119

012023
OPEN ACCESS
Plane strain bending under tension of a functionally graded sheet at large strains as an ideal flow process

Sergei Alexandrov and Alexander Pirumov.....124

012024
OPEN ACCESS
Numerical analysis of heat transfer in unsteady nanofluids in a small pipe with pulse pressure

Cheol Park, KiTae Yu, HeeGeun Song, Sedong Kim and Hyomin Jeong.....130

IOP Conf. Series: Journal of Physics: Conf. Series **885** (2017) 011001 doi:10.1088/1742-6596/885/1/011001

PREFACE

It is our great pleasure to welcome you to 2017 The 6th International Conference on Manufacturing Engineering and Process (ICMEP 2017) which will be held in University Lisbon, Portugal, May 27-29, 2017. ICMEP 2017 is dedicated to issues related to Manufacturing Engineering and Process.

The major goal and feature of the conference is to bring academic scientists, engineers, industry researchers together to exchange and share their experiences and research results, and discuss the practical challenges encountered and the solutions adopted. Professors from Portugal, and Turkey are invited to deliver keynote speeches regarding latest information in their respective expertise areas. It will be a golden opportunity for the students, researchers and engineers to interact with the experts and specialists to get their advice or consultation on technical matters, sales and marketing strategies.

This proceedings present a selection from papers submitted to the conference from universities, research institutes and industries. All of the papers were subjected to peer-review by conference committee members and international reviewers. The papers selected depended on their quality and their relevancy to the conference. The volume tends to present to the readers the recent advances in the field of Manufacturing Engineering and Process and various related areas, such as Biological materials and processing technology, Analysis and application of material properties, Mechanical engineering and computer aided design, Manufacturing systems and production management, Experiment and analysis of mechanical properties of functional materials, etc.

We would like to thank all the authors who have contributed to this volume and also to the organizing committee, reviewers, speakers, chairpersons, sponsors and all the conference participants for their support to ICMEP 2017.

Prof. Mário S. Ming Kong
University Lisbon, Portugal
June 1, 2017.

Conference Committee

International Advisory Committee

João Seixas, Superior Technical Institute of Lisbon (IST/UL),University Lisbon, Portugal

General Chair

Mário S. Ming Kong, University Lisbon, Portugal

Osman Adiguzel, Firat University, Department of Physics, Turkey

Program Chairs

Andrea Deaconescu, Transilvania University of Brasov, Romania

İlhan Asiltürk, Selcuk University, Turkey

Serhat İkizoğlu, Istanbul Technical University, Turkey

Technical Committee

Joerg Niemann, University of Applied Sciences Düsseldorf, Germany

Cherkaoui Abdelghani, MOHAMED V UNIVERSITY, EMI (Mohammadia School of Engineers), Morocco

Prof. Dr.Porojan Liliana,"V. Babeş" University of Medicine and Pharmacy, Timişoara, Romania

Hatice S. Sas, Abdullah Gul University, Turkey

Francisco Klebson Gomes dos Santos, Universidade Federal Rural do Semi Arido, Brazil

Mohammed Sriti, Engineering Sciences laboratory, Polydisciplinary Faculty of Taza, Sidi Mohamed Ben Abdellah University, Morocco

Gastão Coelho de Aquino Filho, Federal Institute of Education, Science and Technology of Paraíba IFPB-Campus Cajazeiras, Brasil

Thien-My, DAO, Ecole de technologie Supérieure/University of Québec Montreal, Québec, Canada

José B. Aguiar, University of Minho,Guimataes,Portugal

Halit S. TURKMEN, Istanbul Technical University, Turkey

BALAN GEORGE, University of Sibiu, Romania

Yahya Bozkurt, Marmara University, Turkey

Vladimir Strezov, Macquarie University, Australia

Sureerat Polsilapa, Kasetsart University, Thailand

Moraru Roland, University of Petrosani, Romania

Muhammad Fahad, NEDUET Karachi Pakistan, Pakistan

Che Hassan Bin Che Haron, National University Of Malaysia, Malaysia

Jaharah Binti A. Ghani, National University Of Malaysia, Malaysia

Balhassn S. M. Ali, Department of Mechanical Engineering, Bursa Orhangazi University, Turkey

Hassan El-Hofy, Production Engineering Department, Faculty of Engineering, Alexandria University, Egypt

Yusri Yusof, Universiti Tun Hussein Onn Malaysia, Malaysia

Mehmet Karakose, Firat University, Turkey

Hasan AL Dabbas, Philadelphia University, Jordan

Supakorn Pukird, Ubon Ratchathani University, Thailand

Mehmet Savsar, Kuwait University, Kuwait

Peer review statement

All papers published in this volume of *Journal of Physics: Conference Series* have been peer reviewed through processes administered by the proceedings Editors. Reviews were conducted by expert referees to the professional and scientific standards expected of a proceedings journal published by IOP Publishing.

IOP Conf. Series: Journal of Physics: Conf. Series **885** (2017) 012001 doi:10.1088/1742-6596/885/1/012001

Optimization of factors to obtain cassava starch films with improved mechanical properties

Mayra Monteiro[1*], Victor Oliveira[2], Francisco Santos[2], Eduardo Barros Neto[1], Karyn Silva[1], Rayane Silva[1], João Henrique[1], Abimaelle Chibério[3]

[1]Universidade Federal do Rio Grande do Norte, UFRN, Natal – RN, Brazil.
[2]Universidade Federal Rural do Semi-Árido, UFERSA, Mossoró – RN, Brazil.
[3]Universidade NOVA de Lisboa, Lisboa, Portugal.

mayra.kerolly@gmail.com

Abstract. In this study, was investigated the optimization of the factors that significantly influenced the mechanical property improvement of cassava starch films through complete factorial design 2^3. The factors to be analyzed were cassava starch, glycerol and modified clay contents. A regression model was proposed by the factorial analysis, aiming to estimate the condition of the individual factors investigated in the optimum state of the mechanical properties of the biofilm, using the following statistical tool: desirability function and response surface. The response variable that delimits the improvement of the mechanical property of the biofilm is the tensile strength, such improvement is obtained by maximizing the response variable. The factorial analysis showed that the best combination of factor configurations to reach the best response was found to be: with 5g of cassava starch, 10% of glycerol and 5% of modified clay, both percentages in relation to the dry mass of starch used. In addition, the starch biofilm showing the lowest response contained 2g of cassava starch, 0% of modified clay and 30% of glycerol, and was consequently considered the worst biofilm.

1. Introduction

Thermoplastic starch is a biodegradable plastic material biologically based which can be obtained from the melting of the native starch granules extracted from a variety of widely available cultures. Due to the chemical structure of the anhydroglucose repeat unit, the two macromolecules constituting the starch (amylose and hyperbranched amylopectin) are highly hydrophilic [11].

Cassava starch is therefore subject to water absorption and plastification by water molecules. This sensitivity of starch to water makes it poorly competitive for the substitution of thermoplastics derived from petroleum in the production of packaging. Moreover, even if the use of a non-volatile plasticizer in the thermoplastic process of starch, such as glycerol, allows the formulation of soft and non-cracking materials by a "stabilization" of the starch, increased molecular mobility may favor the recrystallization of the starch and progressive embrittlement in its structure [9, 10].

An approach to overcome such limitations is the introduction of modified clay nanoparticles in the presence of a quaternary ammonium salt to reinforce the biopolymer structure, such approach showed good results to the corn starch film that had a marked improvement in its mechanical properties [2, 4]. Based on the above and making use of a complete factorial design 2^3, we investigated the individual and interactive effects of three factors that may affect the optimization of the mechanical properties of cassava starch films.

IOP Conf. Series: Journal of Physics: Conf. Series **885** (2017) 012001 doi:10.1088/1742-6596/885/1/012001

2. Materials and methods

2.1. Ion exchange
The ion exchange was performed according to the methodology proposed by Lee et al. [6]. In the first step was formed a solution of 1% by weight of bentonite clay in deionized water. This solution was kept under stirring and heated at 30 ° C in a thermostat until to obtain swollen slurry. In another Becker, the quaternary ammonium salt was dissolved in 20 mL of ethyl alcohol to form a 0.06 M solution. The two solutions were then mixed and kept under stirring for 12 h at 30 ° C. After stirring, the solution was filtered and washed to remove excess salt. The material retained in the filtration was brought to the oven at 60 ° C and held for 24 h. After drying, they were macerated with mortar and pistil, and then sieved in ABNT No. 200 sieve ($\varphi = 0.074$ mm).

2.2. Preparation of biopolymer film
The film was prepared according to the methodology proposed by Cyras et al. [3]. Initially, 10 solutions were prepared each one containing the cassava starch mass and the percentage of glycerol in relation to the dry mass of the polymer, as shown in Table 2. The mixture was heated at 70°C for 15 min until complete homogenization. Then, 10 solutions were prepared with different percentages of modified bentonite clay, in relation to the polymer dry mass, dispersed in distilled water according to Table 2. Each solution was placed in an ultrasonic bath at room temperature. The solutions of starch and clay were mixed and placed in an ultrasonic bath for 30min to facilitate dissolution. The prepared filmogenic solutions were arranged in rectangular plates and dried at 40°C for 6 hours.

2.3. Statistical analysis
Table 1 shows in quantitative terms the ten experimental tests carried out for the factorial analysis of the tensile strength of cassava starch biofilm. Three variables associated to the formation of the nanocomposite were considered as factors that could influence the improvement of the mechanical properties of the starch film, being these the contents of cassava starch (A), glycerol (B) and modified clay (C). Table 1 too establishes the quantitative analysis of the three factors in terms of the minimum, maximum and central levels.

The tensile strength was determined with the average of three experiments in parallel. The order in which experiments were performed was random to avoid systematic errors. The results were analyzed with Statistica 8 software, and the main effects and interactions among factors were determined. The central point was adopted to verify whether or not there is a lack of fit in the empirical model, which would be impossible by analyzing only two levels. The repeats at the central point have two purposes: to provide a measure of the pure error and to stabilize the expected response variance [1, 5].

Table 1. Shows the matrix of the contrast coefficients of factorial design 2^3.

Test	Real Value			Response
	A	B	C	Tensile Strength (MPa)
1	2	10	0	82.873
2	5	10	0	87.235
3	2	30	0	74.793
4	5	30	0	78.729
5	2	10	5	131.412
6	5	10	5	138.329
7	2	30	5	118.600
8	5	30	5	124.842
9	3.5	20	2.5	104.602
10	3.5	20	2.5	104.524

2.4. Characterizations

2.4.1. Visual aspect
At this stage, the biofilms selected showed the best visual aspects, absence of cracks or zones with tendency to rupture, without insoluble particles and uniform color.

2.4.2. Thickness
The thicknesses of the films were measured using a Mitutoyo micrometer (Model MDC-25M, MFG / Japan). The measurements were taken at five different points throughout the film.

2.4.3. Mechanical properties
The mechanical properties were determined using a Testing Machine DL5000 / 10000 Series EMIC 23, EMIC (Parana, Brazil), which operates according to standard ASTM D882-83 method at a test speed of 5 mm / min with application of total force of 5KN. The samples follow the same standard and are evaluated with length of 50mm, width of 5mm and obeying the maximum thickness of 0.25mm [8].

3. Results and discussion

3.1. Factorial analysis

3.1.1. Estimation of optimal design conditions by the desirable function method
Figure 1 shows, according to HANK et al., 2014 [5] the quantitative measurement of each factor to obtain the best and the worst biofilm of cassava starch, based on the regression model proposed by the factorial analysis.

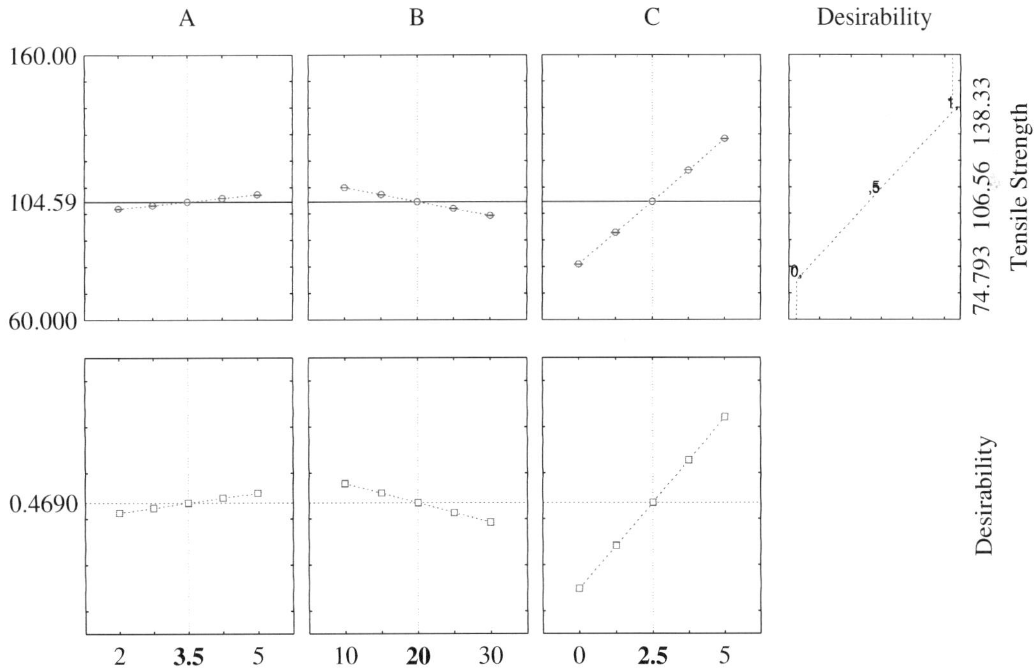

Figure 1. Desirable function for optimizing the response.

The preference function method was used to obtain cassava starch biofilm with the highest tensile strength, that is, with the best mechanical properties. The optimum condition, that is, the best combination of factor configurations to reach the best response was found to be: with 5g of cassava starch, 10% of glycerol and 5% of modified clay, both percentages in relation to the dry mass of starch used. In addition, the biofilm that showed the least tensile strength contained 2 g of cassava starch, 0% of modified clay and 30% of glycerol, and was consequently considered the worst biofilm. The regression model for the factorial analysis performed, which predicts the observed desirability function, is found in [7].

3.1.2. Response surface

Figures 2 show across the response surface that an interaction is effective when the change in response from lows to high levels of a factor is dependent on the level of a second factor, this is, when the lines are not parallel [1,5].

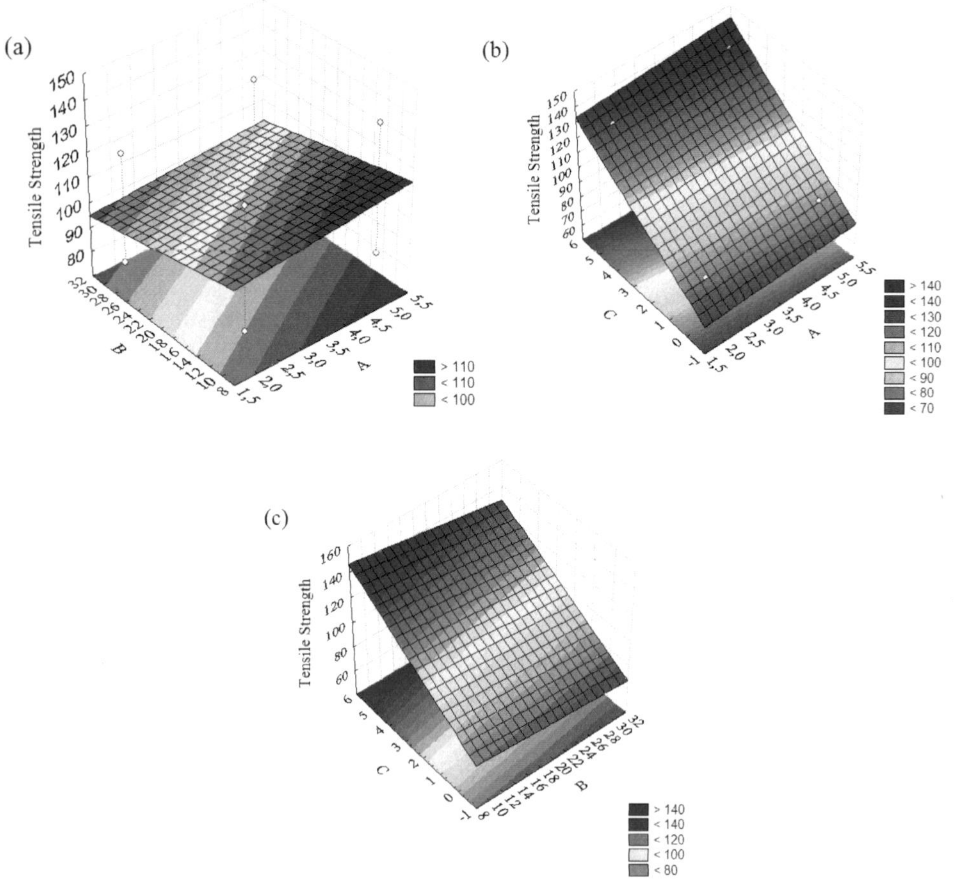

Figure 2. Shows the responses surfaces for the standardized effects: (a) Starch [A] - Glycerol [B], (b) Starch [A] - modified clay [C] and (c) Glycerol [B] - modified clay [C].

The interactions between two factors AB, BC and AC were statistically significant in determining the tensile strength. However, the response surfaces clearly indicated by the slope of the plane representing the statistical model of the observed response variable, that the interaction between B and C was more representative than between A and C, and that the interaction between A and B resulted in the lowest inclination. The AB, BC and AC interactions showed the highest tensile strength at high levels of modified clay, proving that clay is the most significant factor for the improvement of mechanical properties. The effect of C indicated the optimal region of the biofilm strength at the lowest level of A and of B. Therefore, the interaction effects between the AC and BC factors revealed that the nanocomposite with the best mechanical properties were obtained at low A, low B, and high C. These results confirm the observed in the desirability analysis.

4. Conclusion

The statistical design of the experiments combined with regression techniques was applied to optimize the mechanical properties improvement conditions of cassava starch films. The modified clay was the factor of greater statistical significance on the tensile strength of the biofilm observed, being the factor that most contributed to the improvement of the mechanical properties of the biofilm. The factorial experiments showed that the interaction of glycerol with both modified clay and cassava starch was significant for the obtaining of biofilm ductile. The best biofilm contained 5 g of starch, 5% of modified clay and 10% of glycerol; the worst biofilm contained 2 g of starch, 0% of modified clay and 30% of glycerol.

Acknowledgments
Authors thank to Laboratory of Chemical Processes (UFERSA); to Food Technology Laboratory (UFERSA); to Laboratory of Nanostructured Materials and Catalytic Reactors - LAMNRC (UFRN); to Laboratory of Structural Characterization of Materials (UFRN); To Nucleus of Education and Research in Oil and Gas - NUPEG (UFRN).

References

[1] BINGOL, D.; TEKIN, N.; ALKAN, M. Brilliant Yellow dye adsorption onto sepiolite using a full factorial design, Applied Clay Science, v.50, p. 315–321, 2010.

[2] CHIU, C.W.; HUANGB, T.K.; WANGB, Y.C.; ALAMANIB, B.G.; LINB, J.J. Intercalation strategies in clay/polymer hybrids. Progress in Polymer Science, v.39, p. 443–485, 2014.

[3] CYRAS, V. P.; MANFREDI, L. B.; MINH-TAN, T. T.; VAZQUEZ, A. Physical and mechanical properties of thermoplastic starch/montmorillonite nanocomposite films. Carbohydrate Polymers, v. 73, v.55–63, 2008.

[4] GAO, Y.; DAI, Y.; ZHANG, H.; DIAO, E.; HOU, H., DONG, H. Effects of organic modification of montmorillonite on the performance of starch-based nanocomposite films, Applied Clay Science 99, p. 201–206, 2014.

[5] HANK, D.; AZI, Z.; HOCINE, S. A.; CHAALAL, O.; HELLAL, A. Optimization of phenol adsorption onto bentonite by factorial design methodology. Journal of Industrial and Engineering Chemistry, v.20, p.2256–2263, 2014.

[6] LEE, S.; PARK, M.; KIM, D.; Il KIM; PARK, D. Catalytic performance of ion-exchanged montmorillonite with quaternary ammonium salts for the glycerolysis of urea, Catalysis Today, v.232, p.127–133, 2014.

[7] MONTEIRO, M. K. S.; OLIVEIRA, V.R.L.; SANTOS, F.K.G.; NETO, E.L.B.; SILVA, K.N.O.; SILVA, R.R.; HENRIQUE, J.M.M.; CHIBÉRIO, A.S. Statistical analysis of the factors that influenced the mechanical properties improvement of cassava starch films. Journal of Physics: Conference Series, 2017.

[8] ROMERO-BASTIDA, C.A.; BELLO-PÉREZ, L.A.; VELAZQUEZ, G.; ALVAREZ-RAMIREZ, J. Effect of the addition order and amylose content on mechanical, barrier and structural properties of films made with starch and montmorillonite. Carbohydrate Polymers, v.127, p.195–201, 2015.

[9] SOUZA, A.C.; BENZE, R.; FERRÃO, E.S.; DITCHFIELD, C.; COELHO, A.C.V.; TADINI, C.C. Cassava starch biodegradable films: Influence of glycerol and clay nanoparticles content on tensile and barrier properties and glass transition temperature, LWT - Food Science and Technology, v.46, p.110-117, 2012.

[10] VIEIRA, M. G. A.; SILVA, M. A.; SANTOS, L. O.; BEPPU, M. M. Natural-based plasticizers and biopolymer films: A review, European Polymer Journal, v.47, p.254–263, 2011.

[11] ZHU, F. Composition, structure, physicochemical properties, and modifications of cassava starch. Carbohydrate Polymers. v.122, p.456–480, 2015.

Statistical analysis of the factors that influenced the mechanical properties improvement of cassava starch films

Mayra Monteiro[1*], Victor Oliveira[2], Francisco Santos[2], Eduardo Barros Neto[1], Karyn Silva[1], Rayane Silva[1], João Henrique[1], Abimaelle Chibério[3]

[1] Universidade Federal do Rio Grande do Norte, UFRN, Natal – RN, Brazil.
[2] Universidade Federal Rural do Semi-Árido, UFERSA, Mossoró – RN, Brazil.
[3] Universidade NOVA de Lisboa, Lisboa, Portugal.

mayra.kerolly@gmail.com

Abstract. In order to obtain cassava starch films with improved mechanical properties in relation to the synthetic polymer in the packaging production, a complete factorial design 2^3 was carried out in order to investigate which factor significantly influences the tensile strength of the biofilm. The factors to be investigated were cassava starch, glycerol and modified clay contents. Modified bentonite clay was used as a filling material of the biofilm. Glycerol was the plasticizer used to thermoplastify cassava starch. The factorial analysis suggested a regression model capable of predicting the optimal mechanical property of the cassava starch film from the maximization of the tensile strength. The reliability of the regression model was tested by the correlation established with the experimental data through the following statistical analyse: Pareto graph. The modified clay was the factor of greater statistical significance on the observed response variable, being the factor that contributed most to the improvement of the mechanical property of the starch film. The factorial experiments showed that the interaction of glycerol with both modified clay and cassava starch was significant for the reduction of biofilm ductility. Modified clay and cassava starch contributed to the maximization of biofilm ductility, while glycerol contributed to the minimization.

1. Introduction

The wide application of biodegradable polymers in the production of packaging and as a coating of fruits and vegetables has been receiving more attention by numerous researchers in recent years due to its sustainable character [11].

Starch is a natural polymer that is cheap and readily available. However, its high sensitivity to water limits its use in the production of flexible packaging [13]. Recent research focuses on the incorporation of layered silicates along the biopolymer matrix, especially the chemically modified sodium montmorillonite in the presence of organic surfactants, which could be used as a filling material that enhances the ability of the biopolymer structure to withstand a charge applied, the clay thus serves as a barrier to traction [3, 6].

In this perspective, a very low level of modified montmorillonite loading can form nanocomposites with improved mechanical, thermal, electrical and barrier properties in relation to the physicochemical properties of the original biopolymer. The central issue discussed in the whole preparation of nanocomposites is the dispersion of the nanoparticle of clay influenced by the content of the plasticizer material that thermoplasticizes the cassava starch [1, 4].

IOP Conf. Series: Journal of Physics: Conf. Series **885** (2017) 012002 doi:10.1088/1742-6596/885/1/012002

Based on the above and making use of a complete factorial design 2^3, was investigated the individual and interactive effects of three factors that may affect the optimization of the mechanical properties of cassava starch films.

2. Materials and methods

2.1. Ion exchange
The ion exchange was performed according to the methodology proposed by Lee et al. [7]. In the first step was formed a solution of 1% by weight of bentonite clay in deionized water. This solution was kept under stirring and heated at 30°C in a thermostat until to obtain swollen slurry. In another Becker, the quaternary ammonium salt was dissolved in 20 mL of ethyl alcohol to form a 0.06 M solution. The two solutions were then mixed and kept under stirring for 12h at 30°C. After stirring, the solution was filtered and washed to remove excess salt. The material retained in the filtration was brought to the oven at 60°C and held for 24 h. After drying, they were macerated with mortar and pistil, and then sieved in ABNT No. 200 sieve ($\varphi = 0.074$ mm).

2.2. Preparation of biopolymer film
The film was prepared according to the methodology proposed by Cyras et al. [5]. Initially, 10 solutions were prepared each one containing the cassava starch mass and the percentage of glycerol in relation to the dry mass of the polymer, as shown in Table 2. The mixture was heated at 70 ° C for 15 min until complete homogenization. Then, 10 solutions were prepared with different percentages of modified bentonite clay, in relation to the polymer dry mass, dispersed in distilled water according to Table 2. Each solution was placed in an ultrasonic bath at room temperature. The solutions of starch and clay were mixed and placed in an ultrasonic bath for 30min to facilitate dissolution. The prepared filmogenic solutions were arranged in rectangular plates and dried at 40 ° C for 6 hours.

2.3. Statistical analysis
Table 1 shows in quantitative terms the ten experimental tests performed for the factorial analysis of the tensile strength of the cassava starch biofilm, as well as table 2 that establishes the quantitative analysis of the three factors in terms of minimum, maximum and central levels, are shown in [8]. The breaking strength was determined with the average of three experiments in parallel. The order in which experiments were performed was random to avoid systematic errors. The results were analyzed with Statistica 8 software, and the main effects and interactions among factors were determined.

2.4. Characterizations

2.4.1. Visual aspect
At this stage, the biofilms selected showed the best visual aspects, absence of cracks or zones with tendency to rupture, without insoluble particles and uniform color.

2.4.2. Thickness
The thicknesses of the films were measured using a Mitutoyo micrometer (Model MDC-25M, MFG / Japan). The measurements were taken at five different points throughout the film.

2.4.3. Mechanical properties
The mechanical properties were determined using a Testing Machine DL5000 / 10000 Series EMIC 23, EMIC (Parana, Brazil), which operates according to standard ASTM D882-83 method at a test speed of 5 mm / min with application of total force of 5KN. The samples follow the same standard and are evaluated with length of 50mm, width of 5mm and obeying the maximum thickness of 0.25mm [5].

3. Results and discussion

IOP Conf. Series: Journal of Physics: Conf. Series **885** (2017) 012002 doi:10.1088/1742-6596/885/1/012002

3.1. Factorial analysis

Selected the three important factors, the next step was to quantitatively evaluate their influence on the response of interest, with the minimum of experiments using a full factorial design 2^3. Therefore, the effect of one factor depends on the level of the other, and it is possible to calculate the value of the interaction effect between them. In terms of the effects were obtained the correlation coefficients of the statistical model used to describe the response of the complete factorial planning. The T-value indicates the contribution of each effect on the observed response. The P-value indicates that the observed result has more than 95% confidence. Table 1 shows all the statistical parameters cited above having the tensile strength as a response variable.

Table 1. Shows the estimated effects of the three investigated factors, the correlation coefficients of the model and the statistical parameters for the tensile strength response.

Parameters	Effects	Coefficients	Default Error	T – Value	P – Value
Mean	104.594	104.5937	0.016367	6390.429	0.00E+00
A	5.3642	2.6821	0.036598	146.569	0.000047
B	-10.721	-5.36065	0.036598	-292.946	0.000012
C	47.3882	23.6941	0.036598	1294.82	0.000001
A – B	-0.2749	-0.13745	0.036598	-7.511	0.017266
A – C	1.2151	0.60755	0.036598	33.201	0.000906
B – C	-2.4286	-1.2143	0.036598	-66.357	0.000227
A – B – C	-0.0623	-0.0312	0.036598	-1.701	0.230961

The effects shown in Table 1 can be explained taking into account that the contribution of each factor and the interaction between them on the response obtained was investigated. In this sense, it was observed that both the individual and the interaction effects of cassava starch (A) and glycerol (B) significantly influenced the biofilm tensile strength. The cassava starch contributes to the increase of the tensile strength due to its elasticity from the extensive monomeric chain formed by hydrocarbons and branched by hydroxyl groups from the predominance of amylopectin in its composition [13]. Glycerol contributes to reduce the ductility of the biofilm from the thermoplasticization. This is due to its highly hydrophilic character, which is responsible for reducing the forces that keep the biopolymer matrix cohesive, by breaking the internal hydrogen bonds and the forces of inter and intramolecular attraction between the polymer chains, while increasing the intermolecular space resulting in the fragility of the structure [1]. As a result, the interaction between A and B was also significant for the reduction of film tensile strength.

In this perspective, the individual effects of the three factors were significant on the observed response, with the modified clay (C) contributing the most to the increase. This can be explained because the clay presents a stacked structure of layers interconnected by Van der Waals forces in the presence of free inorganic cations in the interlamellar galleries. Therefore, it becomes organophilized by the ion exchange of inorganic sodium cations by organic cationic surfactants. It has an extensive carbonic chain responsible for increasing the basal distance of the silicate layers, according to [3], [6] and [12], this increase promotes to the thermoplastic biopolymer matrix space for its interlacing that disperses the clay evenly throughout its structure due to compatibility with the functional group of the organic surfactant, thus resulting in a ductile nanocomposite and consequently with the mechanical property of the original biofilm improved. Indeed, in terms of the interaction between two factors, AC interaction also contributed significantly to increased response.

The BC interaction contributed with a negative effect on the resistance of the biofilm, since the glycerol to be depended on the concentration will bind by difference of electronegativity with the crystalline structure of the cassava starch and will make difficult the adequate aggregation of the modified clay [1, 4]. The interaction between the three factors was neglected because they did not significantly influence the ductility of the biofilm, a similar result was found by CHIVRAC (2010) [4] where it is argued that the result is obtained from the deposition of the plasticizer on the starch occurring before the modified clay, this fact leads to the formation of hydrogen bonds between the cassava starch and the glycerol that diminish the attractive forces between the starch and the clay. The mathematical model coded for a complete factorial design 2^3 is represented by:

$$Y = 104.5937 + 2.6821X_1 - 5.3606X_2 + 23.6941X_3 - 0.1374X_1X_2 + 0.6075X_1X_3 - 1.2143X_2X_3 - 0.0312X_1X_2X_3 \quad (1)$$

3.2. Pareto graph

Figure 1 shows the Pareto graph showing which of the individual factors analyzed, as well as which interactions between these factors contribute significantly in the optimization of the mechanical properties of cassava starch biofilms.

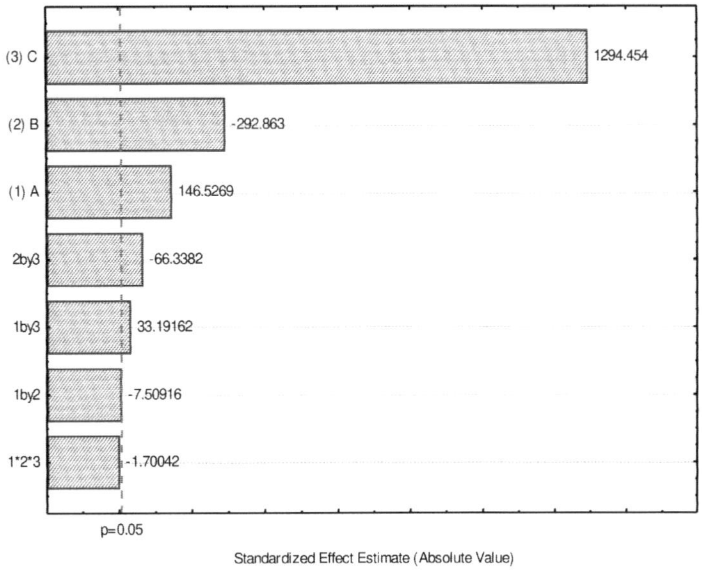

Figure 1. Pareto graph of standardized effects.

For the 95% confidence level and 2 degrees of freedom, the t-critical is equal to 2.920. The value of the t-critical estimated by the student t test was used as the reference line to determine if the calculated effects were significantly different from zero, these values for each effect are shown in the Pareto graph by horizontal columns according to Fig.1, but they are not significant factors [2,9,10]. However, the main factors (A, B and C) and their interactions (AB, AC and BC) that touched the reference line at the 0.05 level were considered significant. The Pareto graph also shows that C represents the individual effect of a greater significant load on the tensile strength, in relation to the interaction between the two factors is BC that presents this characteristic. In addition, only the interaction between the three ABC factors was not significant for the observed response at 95% confidence. The sequence of the degree of statistical significance on the tensile strength in relation to the effects investigated is therefore C> B> A> BC> AC> AB. The individual factors A and C, as well as the interaction between them, AC, show positive effects and contribute to the maximization of the tensile strength of the biofilm. The effects B, AB and BC presented negative effects, leading to the minimization of the response variable. The observed in the Pareto graph confirms the one evidenced in the factorial analysis.

4. Conclusion

The statistical design of the experiments combined with regression techniques was applied to optimize the mechanical properties improvement conditions of cassava starch films. The modified clay was the factor of greater statistical significance on the tensile strength of the biofilm observed, being the factor that most contributed to the improvement of the mechanical properties of the biofilm. The factorial experiments showed that the interaction of glycerol with both modified clay and cassava starch was significant for the reduction of biofilm ductility. The evidenced in the factorial analysis was correlated satisfactorily with that observed in the Pareto graph.

References

[1] ALVES, V.D.; MALI, S.; BELÉIA, A.; GROSSMANN, M.V.E. Effect of glycerol and amylose enrichment on cassava starch film properties, Journal of Food Engineering, v.78, p. 941–946, 2007.
[2] BINGOL, D.; TEKIN, N.; ALKAN, M. Brilliant Yellow dye adsorption onto sepiolite using a full factorial design, Applied Clay Science, v.50, p. 315–321, 2010.
[3] CHIU, C.W.; HUANGB, T.K.; WANGB, Y.C.; ALAMANIB, B.G.; LINB, J.J. Intercalation strategies in clay/polymer hybrids. Progress in Polymer Science, v.39, p. 443–485, 2014.
[4] CHIVRAC, F.; POLLET, E.; DOLE, P.; AVÉROUS, L. Starch-based nano-biocomposites: Plasticizer impact on the montmorillonite exfoliation process, Carbohydrate Polymers, v.79, p.941–947, 2010.
[5] CYRAS, V. P.; MANFREDI, L. B.; MINH-TAN, T. T.; VAZQUEZ, A. Physical and mechanical properties of thermoplastic starch/montmorillonite nanocomposite films. Carbohydrate Polymers, v. 73, v.55–63, 2008.
[6] KOTAL, M.; BHOWMICK, A. K. Polymer nanocomposites from modified clays: Recent advances and challenges. Progress in Polymer Science, v.51, p. 127–187, 2015.
[7] LEE, S.; PARK, M.; KIM, D.; II KIM; PARK, D. Catalytic performance of ion-exchanged montmorillonite with quaternary ammonium salts for the glycerolysis of urea, Catalysis Today, v.232, p.127–133, 2014.
[8] MONTEIRO, M. K. S.; OLIVEIRA, V.R.L.; SANTOS, F.K.G.; NETO, E.L.B.; SILVA, K.N.O.; SILVA, R.R.; HENRIQUE, J.M.M.; CHIBÉRIO, A.S. Optimization of factors to obtain cassava starch films with improved mechanical properties. Journal of Physics: Conference Series, 2017.
[9] PALANIKUMARA, K.; DAVIM, J. P. Assessment of some factors influencing tool wear on the machining of glass fibre-reinforced plastics by coated cemented carbide tools, journal of materials processing technology, v.2 0 9, p.511–519, 2009.
[10] PONNUSAMI, V.; KRITHIKA, V.; MADHURAM, R.; SRIVASTAVA, S.N. Biosorption of reactive dye using acid-treated rice husk: Factorial design analysis, Journal of Hazardous Materials, v.142, p.397–403, 2007.
[11] REDDY, M.M.; VIVEKANANDHAN, S.; MISRA, M.; BHATIA, S. K.; MOHANTY, A.K. Biobased plastics and bionanocomposites: Current status and future opportunities, Progress in Polymer Science, v.38, p.1653– 1689, 2013.
[12] ROMERO-BASTIDA, C.A.; BELLO-PÉREZ, L.A.; VELAZQUEZ, G.; ALVAREZ-RAMIREZ, J. Effect of the addition order and amylose content on mechanical, barrier and structural properties of films made with starch and montmorillonite. Carbohydrate Polymers, v.127, p.195–201, 2015.
[13] ZHU, F. Composition, structure, physicochemical properties, and modifications of cassava starch. Carbohydrate Polymers. v.122, p.456–480, 2015.

IOP Conf. Series: Journal of Physics: Conf. Series **885** (2017) 012003 doi:10.1088/1742-6596/885/1/012003

Failure analysis of various monolithic posterior aesthetic dental crowns using finite element method

Liliana Porojan*[a], Florin Topală[b]

aDepartment of Dental Prostheses Technology, School of Dentistry, "V. Babeş" University of Medicine and Pharmacy, 9 Revolutiei 1989 Blv., 300041 Timişoara, Romania;
bDepartment of Dental Prosthodontics, School of Dentistry, "V. Babeş" University of Medicine and Pharmacy, 9 Revolutiei 1989 Blv., 300041 Timişoara, Romania

* lilianasandu@gmail.com

Abstract. The aim of the study was to assess the effect of material stiffness and load on the biomechanical performance of the monolithic full-coverage posterior aesthetic dental crowns using finite element analysis. Three restorative materials for monolithic dental crowns were selected for the study: zirconia; lithium disilicate glass-ceramic, and resin-based composite. Stresses were calculated in the crowns for all materials and in the teeth structures, under different load values. The experiments show that dental crowns made from all this new aesthetic materials processed by CAD/CAM technologies would be indicated as monolithic dental crowns for posterior areas.

1. Introduction

Due to the increased demand for restorations that look like natural teeth, monolithic ceramic and composite crowns are increasingly used. CAD/CAM technologies have become more widespread to process dental materials in recent years, because of the fact that classical technologies usually require a very long processing time in a well-controlled environment. Ceramic crowns, like zirconia and glass-ceramic, are biocompatible and aesthetic but they are brittle, and failure from fracture continues to be a major concern particularly on posterior teeth which are subjected to greater masticatory loads and more continuous function1. Several high strength ceramic materials have become available and are being used in dental practices [2-4].

In addition to the ceramic blocks, which are more frequently used for CAD/CAM applications, composite resin-based blocks have been developed, with better mechanical properties [5]. There are also some advantages of resin composite blocks over conventional ceramic blocks. They are more easily adjusted, milled and repaired. Micro-tensile bond strength testing showed that resin composite has higher bond strength to resin-based adhesive materials than to glass–ceramics [6]. It was also been found that resin composite caused a significantly lower volume loss on antagonist enamel than ceramics did in two-body wear testing, which may help to preserve the functional balance of dentition [7].

These materials are often introduced into the market without a basic understanding of their clinical performance because long term controlled clinical trials are not required and are both time-consuming and expensive. The development of experimental methods could be useful to help predict clinical behaviour of these new restorative materials1.

Finite element analysis (FEA) is a powerful and flexible computational tool to model dental structures and devices, simulate the occlusal loading conditions and predict the stress and strain distribution. It has been widely used in the dental and medical fields since the 1970s and has been proven to be accurate and efficient for finding solutions for complex geometry problems [8,9].

Other paragraphs are indented (BodytextIndented style).

2. Purpose

The aim of the study was to construct finite element (FE) models for monolithic full-coverage dental crowns for molars and premolars, comparing the stress distribution between zirconia, glass ceramic and composite materials, and revealing the effect of material stiffness and load on the biomechanical performance of the dental crowns.

3. Experimental part

For the experimental analyses a maxillary right first premolar and a maxillary right first molar were chosen. They were designed with a preparation for full-coverage crowns according to the guidelines of the prosthetic material manufacturers. The preparation included a 1.5-mm occlusal reduction, a 1.0 mm axial reduction, a 6-degree angle of taper, a 1-mm circular chamfer shoulder. Three restorative materials for monolithic dental crowns were selected for the study, each in the form of commercially available CAD/CAM blocks: zirconia; lithium disilicate glass-ceramic; and resin-based composite with ceramic filler.

Geometric models of the monolithic crowns were designed to occupy the space between the original tooth form and the prepared tooth form. At first nonparametric modeling software (Blender 2.57b) was used to obtain the 3D tooth shapes. The collected data were used to construct three dimensional models using Rhinoceros (McNeel North America) NURBS (Nonuniform Rational B-Splines) modeling program. The geometric models were imported in the finite element analysis software ANSYS, meshed and finite element calculations were carried out. The finite element model was solved under a variety of load case-material property combinations (Figure 1).

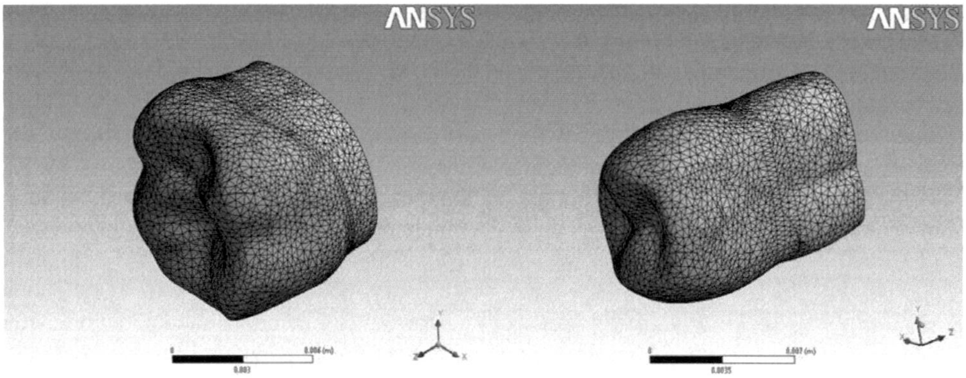

Figure 1. Finite element models of the molar and premolar.

The nodes on the bottom surface of the abutment tooth were constrained in all directions. The crown materials were assigned using literature data. In order to simulate the stress distribution, the Young's module and Poisson's ratios were introduced: Young's modulus (GPa) 18 for dentin, 16 for composite, 64 for glass ceramic, and 205 for zirconia and Poisson's ratio 0.27 for dentin, 0.24 for composite, 0.21 for glass ceramic, and 0.31 for zirconia. The following loading situations were examined respectively: load applied at an average human bite force of 200 N and a maximum bite force (600 N for premolars and 800 N for molars). The simulated bite force was distributed across contact areas with antagonists.

The distribution of the principal stress and the equivalent stress, and also the maximal values were determined.

4. Results and discussions

Stresses were calculated in the crowns for all materials and in the teeth structures, under different load values (Table 1-4, Figures 2,3).

Table1. Maximal principal stresses in the molar crowns and dentin for a 200 N loading

Maximal principal stress values [Pa]	Crown	Dentin
Zirconia	3.23x107	2.02x106
Glass ceramic	4.07x107	1.69x106
Composite	4.16 x107	2.72x106

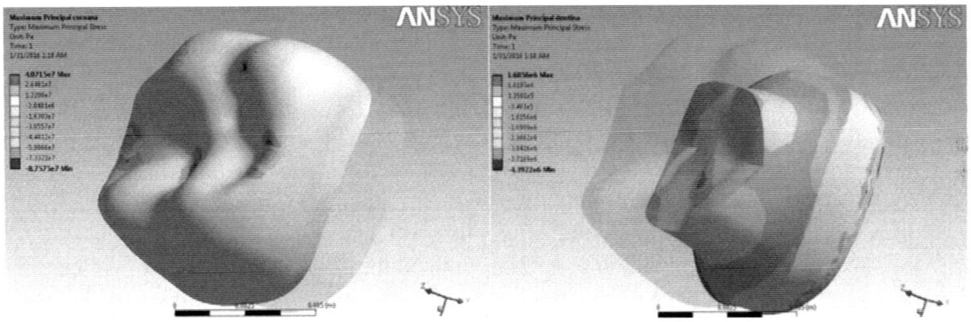

Figure 2. Principal stress distribution in the glass-ceramic molar crown and underlying dentin.

Table 2. Maximal principal stresses in the premolar crowns and dentin for a 200 N loading.

Maximal principal stress values [Pa]	Crown	Dentin
Zirconia	2.53x107	13.4x106
Glass ceramic	2.81x107	13x106
Composite	2.76 x107	12x106

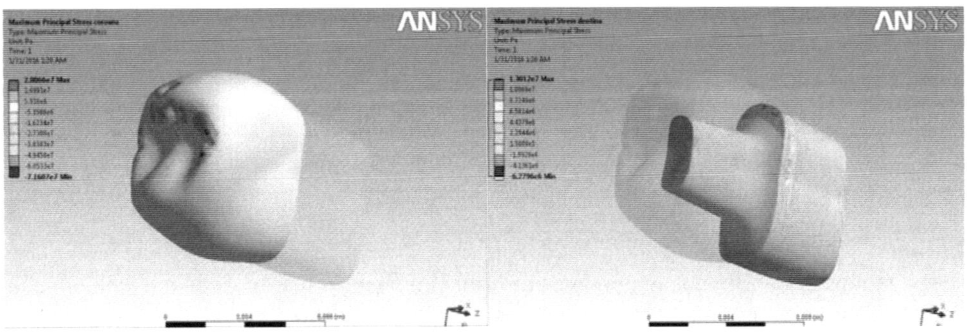

Figure 3. Principal stress distribution in the glass-ceramic premolar crown and underlying dentin.

Table 3. Maximal principal stresses in the molar crowns and dentin for a 800 N loading.

Maximal principal stress values [Pa]	Crown	Dentin
Zirconia	12.9x107	8.07x106
Glass ceramic	16.3x107	6.74x106
Composite	16.6 x107	10.9x106

Table 4. Maximal principal stresses in the premolar crowns and dentin for a 600 N loading.

Maximal principal stress values [Pa]	Crown	Dentin
Zirconia	7.58x107	40.1x106
Glass ceramic	8.42x107	39x106
Composite	8.28x107	35.9x106

Compared to the tensile strength of the new aesthetic materials, 745 MPa for zirconia, and 360 MPa for glass ceramic, and 200 MPa for composite, the maximal principal stresses in the crowns don't exceed them for 600 N in the premolar area and 800 N load in the molar area, for all type of materials. According to the literature data [10], that the tensile strengths of dentin ranged from 44.4 MPa to 97.8 MPa, no harmful effects occur in hard teeth structures, because in all cases first principal stresses in dentin are lower. The values approach the lower limit in case of premolar crowns and result in increased stress levels, eventually leading to a compromised long-term clinical performance.

The material is important to withstand increased loads which occur during functions. Reported loads during normal function vary considerably; there is no accepted consensus on either loads present in vivo or how they should be replicated in vitro. Some authors use lower loads, 100-200 N, others use loads in the range of 500-800 N [11].

It is not surprising that attempts would be made to substitute highly filled composite CAD/CAM blocks for ceramic blocks in fabricating full-coverage crowns [5]. In vitro testing studies compared CAD/CAM composite to ceramics favorably in terms of monotonic failure load and showed performance similar to veneering ceramics in fatigue loading [12-15]. However, dental composites have a discontinuous ceramic phase, and so are limited to much lower elastic modulus values than glass–ceramic materials. It has been found that thin-walled crowns made of low elastic modulus material (composite resin) are more prone to debonding than those made of stiffer materials such as ceramics. While thin-walled crowns of stiffer materials can protect tooth structures from damage better than composite resin ones. This lack of stiffness may result in increased stress levels under the same bite force compared to glass–ceramic restorations, eventually leading to a compromised long-term clinical performance.

Further investigations are still needed to analyze and evaluate the biomechanical behavior of this new type of aesthetic materials designed for dental crowns, relative to the material properties, processing technology and of course to the restoration design.

5. Conclusions

Within the limitations of the present study, the following conclusions can be drawn:

1. The distribution of principal stresses was not dependent on the bite force.
2. Maximal values of the principal stresses didn't exceed the tensile strength values of the new aesthetic materials with improved mechanical properties.
3. Principal stresses were located at the contact areas with the antagonists and these areas can be starting points for the failure of aesthetic crowns in the posterior area.
4. Dental crowns made from all this new aesthetic materials processed by CAD/CAM technologies would be indicated as monolithic dental crowns for the posterior areas.

Acknowledgments
This work was supported by a grant of the Romanian National Authority for Scientific Research and Innovation CNCS-UEFISCDI, project number PN-II-RU-TE-2014-4-0476.

References

[1] Nasrin, S., Katsube, N, Seghi, R.R., Rokhlin, S.I., "3D statistical failure analysis of monolithic dental ceramic crowns", J. Biomech., , 49(10), 2038-46 (2016).
[2] Della Bona, A., Mecholsky, J.J., Anusavice, K.J., "Fracture behavior of lithia disilicate- and leucite-based ceramics. Fracture behavior of lithia disilicate- and leucite-based ceramics", Dent. Mater., 20, 956-62 (2004).
[3] Denry, I., Kelly, J.R., "State of the art of zirconia for dental applications", Dent. Mater. 24, 299–307 (2008).
[4] Sandu, L., Topala, F., Porojan, S., "Marginal Design Evaluation for CAM Obtained Zirconia Based Crown Frameworks", Advanced Materials Research, 213, 349-353, (2011).
[5] Duan, Y, Grigg, J.A., "Effect of elasticity on stress distribution in CAD/CAM dental crowns: Glass ceramic vs. polymer-matrix composite", J. Dent., 43(6)742-9 (2015).
[6] El Zohairy, A.A., De Gee, A.J., Mohsen, M.M., Feilzer, A.J., "Microtensile bond strength testing of luting cements to prefabricated CAD/CAM ceramic and composite blocks", Dental Materials, 19,575-83 (2003).
[7] http://multimedia.3m.com/mws/media/781394O/lava-ultimate-technical-product-profile-global.pdf. (2011)
[8] Savoldelli, C., Tillier, Y., Bouchard, P.O., Odin, G., "Contribution of the finite element method in maxillofacial surgery", Revue de Stomatologie et de Chirurgie Maxillo-faciale, 110,27-33 (2009).
[9] Van Staden, R.C., Guan, H., Loo, Y.C., "Application of the finite element method in dental implant research. Computational Methods" Biomechanics and Biomedical Engineering, 9,257–70 (2006).
[10] Staninec. M., Marshall, G.W., Hilton, J.F., Pashley, D.H., Gansky, S.A., Marshall, S.J., Kinney, J.H., "Ultimate tensile strength of dentin: Evidence for a damage mechanics approach to dentin failure". J. Biomed. Mater. Res., 63(3),342-5 (2002).
[11] Kohorst, P., Dittmer, M.P., Borchers, L., Stiesch-Scholz, M., "Influence of cyclic fatigue in water on the load-bearing capacity of dental bridges made of zirconia", Acta Biomater., 4(5)1440-7 (2008).
[12] Chen, C., Trindade, F.Z., De Jager, N., Kleverlaan, C.J., Feilzer, A.J., "The fracture resistance of a CAD/CAM Resin Nano Ceramic (RNC) and a CAD ceramic at different thicknesses", Dental Materials, 30,954-62, (2014).
[13] El-Damanhoury, H., Haj-Ali, R., Platt, J., "Fracture resistance and microleakage of endocrowns utilizing three CAD-CAM blocks", Operative Dentistry, 40(2):201-10 (2015).
[14] Ilgenstein, I., Zitzmann, N.U., Buhler, J., Wegehaupt, F.J., Attin, T., Weiger, R., Krastl, G., "Influence of proximal box elevation on the marginal quality and fracture behavior of root-filled molars restored with CAD/CAM ceramic or composite onlays", Clinical Oral Investigations, 19,5,1021-8 (2015).
[15] Belli, R., Geinzer, E., Muschweck, A., Petschelt, A., Lohbauer, U., "Mechanical fatigue degradation of ceramics versus resin composites for dental restorations", Dental Materials, 30,424-32 (2014).

IOP Conf. Series: Journal of Physics: Conf. Series **885** (2017) 012004 doi:10.1088/1742-6596/885/1/012004

Surface texture and hardness of dental alloys processed by alternative technologies

Liliana Porojan*[a], Cristina E. Savencu[a], Florin I. Topală[b], Sorin D. Porojan[c]

[a]Department of Dental Prostheses Technology, Faculty of Dentistry, "V. Babeş" University of Medicine and Pharmacy, 9 Revolutiei 1989 Blv., 300041 Timişoara, Romania;
[b]Department of Dental Prosthodontics, School of Dentistry, "V. Babeş" University of Medicine and Pharmacy, 9 Revolutiei 1989 Blv., 300041 Timişoara, Romania;
[c]Department of Oral Rehabilitation, Faculty of Dentistry, "V. Babeş" University of Medicine and Pharmacy, 9 Revolutiei 1989 Blv., 300041 Timişoara, Romania

*lilianasandu@gmail.com; phone 0040722310299

Abstract. Technological developments have led to the implementation of novel digitalized manufacturing methods for the production of metallic structures in prosthetic dentistry. These technologies can be classified as based on subtractive manufacturing, assisted by computer-aided design/computer-aided manufacturing (CAD/CAM) systems, or on additive manufacturing (AM), such as the recently developed laser-based methods. The aim of the study was to assess the surface texture and hardness of metallic structures for dental restorations obtained by alternative technologies: conventional casting (CST), computerized milling (MIL), AM power bed fusion methods, respective selective laser melting (SLM) and selective laser sintering (SLS). For the experimental analyses metallic specimens made of Co-Cr dental alloys were prepared as indicated by the manufacturers. The specimen structure at the macro level was observed by an optical microscope and micro-hardness was measured in all substrates. Metallic frameworks obtained by AM are characterized by increased hardness, depending also on the surface processing. The formation of microstructural defects can be better controlled and avoided during SLM and MIL process. Application of power bed fusion techniques, like SLS and SLM, is currently a challenge in dental alloys processing.

1. Introduction

Alternative manufacturing methods of dental restorations using base metal dental alloys are of ongoing interest, having a growing impact in the field of dental technology.

Technological developments have led to the implementation of novel digitalized manufacturing methods for the production of metallic structures in prosthetic dentistry [1-5]. The potential for fabricating metallic dental components with a complex geometry directly from digital data using automated equipment and appropriate materials is very significant. These technologies can be classified as based on subtractive manufacturing, such as the milling of premanufactured materials assisted by computer-aided design/computer-aided manufacturing (CAD/CAM) systems [2,5-7], or on additive manufacturing (AM), such as the recently developed laser-based methods. Although CAD/CAM has long been directly associated with the milling procedure in dental literature, it should be mentioned that AM procedures are also classified as CAD/CAM technologies.

Understanding of structural and microstructural defects present in metallic frameworks processed through AM techniques and post-processing protocols such as heat treatment to reduce these defects, are important in order to transform the microstructures to those acceptable for practice [8].
Selective laser melting (SLM) and Selective laser sintering (SLS) belong to Powder bed fusion, and consist in a thermal energy that selectively fuses regions of a powder bed [9]. Direct metal printing methods can generally be categorized as laser-based, electron beam-based, arc-based, and ultrasonic welding-based. Laser-based metal AM methods are classified into laser sintering (SLS), laser melting (SLM), and laser metal deposition (LMD) [10].
SLS is a powder bed fusion technique in which a scanning laser is used to consolidate sequentially deposited layers of a metal powder [11]. Different types of lasers including CO2, disk, Nd:YAG, and fiber lasers are used. The principal consolidation mechanism is liquid-phase sintering involving partial melting and coalescence of the powder. SLM is a second powder bed fusion technique involving the consolidation of metal powders using powerful lasers. While the equipment setup and configuration and processing methodology are similar in SLS and SLM, in SLM, the powder is completely or nearly completely melted to produce a fully dense or nearly fully dense structure. SLM thus produces metal articles with a higher level of microstructural homogeneity compared with SLS.

2. Objective

The aim of the study was to assess the surface texture and hardness of metallic structures for dental restorations obtained by alternative technologies: conventional casting, computerized milling, AM power bed fusion methods, respective SLS and SLM.

3. Materials and methods

For the experimental analyses metallic specimens made of Co-Cr dental alloys were prepared using traditional casting (CST), computerized milling (MIL), selective laser sintering (SLS) and selective laser melting (SLM), as indicated by the manufacturers. Round plates of 20 mm diameter and 2 mm thick were fabricated using different technologies. The cast alloy was manufactured by conventional lost wax technique with a phosphate-bonded investment. The mold was cast with Wirobond® SG alloy (Bego, Bremen, Germany) at 900°C using a vacuum pressure casting machines Nautilus (Bego, Bremen, Germany). The mold was left to cool down to room temperature and the specimens were then divested, sandblasted with alumina particles (200 μm) and finished with rotative instruments, burs suitable for Co-Cr alloys. Type 4 alloy Wirobond® SG fulfils all criteria of the ISO 22674 and ISO 9693-1 Standards. The composition is Co 63.8; Cr 24.8; W 5.3; Mo 5.1; other constituents Si, and Fe. Wirobond® SG allows normal cooling and provides effective, economic processing in the laboratory.
A prefabricated block of a commercial Co-Cr dental alloy CopraBond K (White Peaks Dental Systems GmbH & Co. KG, Essen, Germany), a non precious blank on Co-Cr base, type 4, was milled to fabricate the sample using Datron D5 5-axis dental milling machine (Datron, Mühltal, Germany). A disc wax pattern was digitized and the specimens were cut to their final dimensions. The composition is: Co 61.0, Cr 28.0, W 8.5, Mn 0.25, Fe <0.5, Si 1.65, C <0.1. CopraBond K is a nickel- and beryllium free Co-Cr blank, specially designed for CAD/CAM applications.
The laser-sintered and laser-melted specimens were prepared from commercial Co-Cr powder Starbond CoS Powder 16 (S & S Scheftner GmbH, Mainz, Germany) using a dental laser sintering device (PXS Dental System, Phenix Systems, Clermont-Ferrard, France), respective a SLM device MYSINT 100 (Sisma, Piovene, Italy). For scanning the high accuracy D700 scanner (3Shape, Copenhagen, Denmark) and for design Dental System™ CAD Software (3Shape, Copenhagen, Denmark) was used. The alloy is a type 4 alloy according to ISO 22674. Free of beryllium, nickel and cadmium. Nominal values of alloy composition in mass percent: Co 59.0; Cr 25.0; W 9.5; Mo 3.5; Si max. 1; other constituents: C, Fe, Mn, N max. 1.5.
The high power laser beam fused the powder granules homogeneously distributed on a flat building plate guided by a pre-determined pattern by computer aided design (CAD). The laser power, scanning speed, and the distance between two scanning line sand layers were the principal parameters in the

laser sintering and melting processes. These were set according to two recommendations for dental prostheses, respective metal processing, and using fibre laser for power bed fusion. Relief-firing was conducted under argon by heating up to 450°C within 60 minutes, holding for 45 minutes. The specimens resulted after CAD/CAM technologies were not additional prepared.
Oxide-firing (at 950 – 980°C) was performed, the metal surface was sandblast again with fresh aluminium oxide (approx. 150 μm). All substrates were ground until 2000-grit SiC paper, polished with universal polishing paste (Ivoclar Vivadent AG, Schaan, Principality of Liechtenstein). They were after cleaned in alcohol, rinsed in distilled water and dried with adsorbent paper towels.
The specimen structure at the macro level was observed by an optical microscope Leica DM500 (Leica Microsystems, Wetzlar, Germany) using a reflective mode. Microstructural defects were measured using an image processing program, Image J.
Micro-hardness was measured in all substrates using Mitutoyo SJ 201 device (Mitutoyo Corporation, Kanagawa, Japan). The hardness value was obtained as a mean value of at least 5 indents for samples polished and sandblast with fresh aluminium oxide of approx. 75 μm and approx. 200 μm).

4. Results
Microstructural defects were observed by optical microscopy on well-polished surfaces of Cr-Co-Mo alloy specimens prepared by SLS, SLM, MIL, CST (Fig. 1).

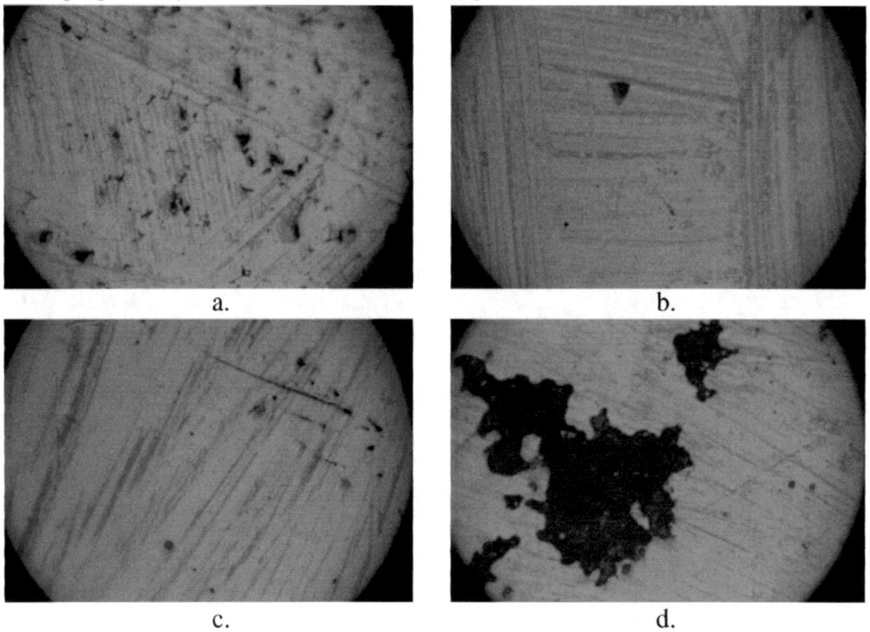

Figure 1. Microstructural defects observed by optical microscopy on well-polished surfaces of Cr-Co-Mo alloy specimens obtained by: a. SLS, b. SLM, c. MIL, d. CST.

Regarding microstructure the best results were obtained for SLM and MIL samples. Isolated or interconnected voids originating from hard agglomerates or insufficient packing of powder granules prior to sintering with sizes up to 45 μm were observed inside the specimens prepared by SLS. The surface of the specimens fabricated by CST contains grooved holes as a result od incomplete melting, with dimensions between 20 and 350 μm. In MIL samples traces were identified which were probably caused during the end-mill process and rare small voids (size up to 20 μm). SLM obtained specimens revealed the best microstructure, with the fewest voids, with dimensions up to 25 μm.
The hardness values recorded for specimens sandblasted with aluminium oxide of 75 μm were the highest for all samples, followed by those sandblasted with aluminium oxide of 200 μm and lastly from the polished (Table 1). Thus the preparation of the frameworks for ceramic veneering using

aluminium oxide of 75 μm is founded also from this point of view. With respect to the manufacturing technique, AM obtained specimens were associated with the highest HV mean values, about two times higher than those processed by MIL and CST .

Table 1. Vickers hardness mean value of the Co-Cr-Mo alloy specimens prepared by SLS, SLM, MIL, CST.

Samples	polished	sandblasted aluminium oxide 75 μm	sandblasted aluminium oxide 200 μm
SLS	431	509	452
SLM	478	506	474
MIL	230	277	273
CST	215	233	231

5. Discussions

Given the large differences in the manufacturing process between casting, which uses the complete melting and overheating of casting materials, the milling of a prefabricated metal block and AM of a fine metallic powder, large differences in microstructural characteristics and hardness values can be anticipated [12].

AM allows obtaining of complex geometries more than subtractive methods. The application of the AM methods, recently introduced in dental technology, and studies using the clinical implications are necessary. By identifying differences in processing methods and material characteristics, future studies should look up for reducing disadvantages [13, 14].

Surface texture metrology can be used as a means of gaining insight into the physical phenomena taking place during the manufacturing process and process variables, through examination of the surface features generated by the process. It becomes a powerful exploration tool, increasing knowledge of the process and ultimately allowing the creation of improved manufacturing processes [15-17]. Surface texture is the geometrical irregularities present at a surface. Surface texture does not include those geometrical irregularities contributing to the form or shape of the surface [18]. The diversity of AM processes as well as the large number of key process parameters that change from build to build makes the use of traditional means of process qualification less than satisfactory [19]. These new challenges require a more profound understanding of the AM technology and process, and will ultimately require the development of AM surface texture good practice guidance, specifications and standards [20].

6. Conclusions

1. The potential for fabricating metallic dental components directly from digital data represents an opportunity.
2. Metallic frameworks obtained by AM are characterized by increased hardness, depending also on the surface processsing.
3. The formation of microstructural defects can be better controlled and avoided during SLM and MIL process.
4. Application of power bed fusion techniques, like SLS and SLM, is currently a challenge in dental alloys processing.

Acknowledgments

This work was supported by a grant of the Romanian National Authority for Scientific Research and Innovation CNCS-UEFISCDI, project number PN-II-RU-TE-2014-4-0476.

References

[1] Al Jabbari YS, Koutsoukis T, Barmpagadaki X, Zinelis S. Metallurgical and interfacial characterization of PFM Co-Cr dental alloys fabricated via casting, milling or selective laser

melting. Dent Mater. 2014 Apr;30(4):e79-88.
[2] van Noort R. The future of dental devices is digital. Dent Mater 2012;28:3–12. Willer J, Rossbach A, Weber HP. Computer-assisted milling of dental restorations using a new CAD/CAM data acquisition system. J Prosthet Dent 1998;80:346–53.
[3] Quante K, Ludwig K, Kern M. Marginal and internal fit of metal–ceramic crowns fabricated with a new laser melting technology. Dent Mater 2008;24:1311–5.
[4] Traini T, Mangano C, Sammons RL, Mangano F, Macchi A, Piattelli A. Direct laser metal sintering as a new approach to fabrication of an isoelastic functionally graded material for manufacture of porous titanium dental implants. Dent Mater 2008;24:1525–33.
[5] Miyazaki T, Hotta Y. CAD/CAM systems available for the fabrication of crown and bridge restorations. Aust Dent J 2011;56(Suppl. 1):97–106.
[6] Willer J, Rossbach A, Weber HP. Computer-assisted milling of dental restorations using a new CAD/CAM data acquisition system. J Prosthet Dent 1998;80:346–53.
[7] Alt V, Hannig M, Wostmann B, Balkenhol M. Fracture strength of temporary fixed partial dentures: CAD/CAM versus directly fabricated restorations. Dent Mater 2011;27:339–47.
[8] Suman Das, David L. Bourell , and S.S. Babu, Metallic materials for 3D printing, MRS Bulletin, vol 41, 2016, p 729-741.
[9] International A: ASTM F2792-12a, Standard Terminology for Additive Manufacturing Technologies. In. West Conshohocken, PA; 2012.
[10] A. Basak , S. Das , Epitaxy and Microstructure Evolution in Metal Additive Manufacturing, Annu. Rev. Mater. Res. 46 , 125 (2016).
[11] Standard Terminology for Additive Manufacturing Technologies (ASTM Stand. F2792–12a, ASTM International, West Conshohocken, PA, 2012)
[12] Al Jabbari YS, Koutsoukis T, Barmpagadaki X, Zinelis S. Metallurgical and interfacial characterization of PFM Co-Cr dental alloys fabricated via casting, milling or selective laser melting. Dent Mater. 2014 Apr;30(4):e79-88.
[13] Bae EJ, Jeong ID, Kim WC, Kim JH. A comparative study of additive and subtractive manufacturing for dental restorations. J Prosthet Dent. 2017 pii: S0022-3913(16)30621-7,
[14] Stephen T. Newman, Zicheng Zhu, Vimal Dhokia, Alborz Shokrani. Process planning for additive and subtractive manufacturing technologies. CIRP Annals -Manufacturing Technology 2015;64:467-70.
[15] A. Townsend, N. Senin, L. Blunt, R.K. Leach, J.S. Taylor, Surface texture metrology for metal additive manufacturing: a review, Precision Engineering, Volume 46, October 2016, Pages 34-47.
[16] Gu D. Laser additive manufacturing of high-performance materials. DE:Springer Verlag; 2015.
[17] Gu D, Shen Y. Balling phenomena in direct laser sintering of stainless steelpowder: metallurgical mechanisms and control methods. Mater Des2009;30:2903–10.
[18] Leach RK. In: Laperrière L, Reinhart G, editors. CIRP encyclopedia ofproduction engineering. Berlin: Springer-Verlag; 2014.
[19] W.E. Frazier, Metal Additive Manufacturing: A Review, J. Mater. Eng. Perform. 23 (6), 1917 (2014).
[20] A. Townsend, N. Senin, L. Blunt, R.K. Leach, J.S. Taylor, Surface texture metrology for metal additive manufacturing: a review, Precision Engineering, Volume 46, October 2016, Pages 34-47.

IOP Conf. Series: Journal of Physics: Conf. Series **885** (2017) 012005 doi:10.1088/1742-6596/885/1/012005

Structural and morphological approach of Co-Cr dental alloys processed by alternative manufacturing technologies

Sorin Porojan[a], Mihaela Bîrdeanu[b], Cristina Savencu[c], Liliana Porojan[c]

[a]Department of Oral Rehabilitation, Faculty of Dentistry, "V. Babeş" University of Medicine and Pharmacy, 9 Revolutiei 1989 Blv., 300041 Timişoara, Romania
[b]National Institute for Research and Development in Electrochemistry and Condensed Matter, 1 Plautius Andronescu Street, 300224 Timisoara, Romania
[c] Department of Dental Prostheses Technology, Faculty of Dentistry, "V. Babeş" University of Medicine and Pharmacy, 9 Revolutiei 1989 Blv., 300041 Timişoara, Romania

lianasandu@gmail.com

Abstract. The integration of digitalized processing technologies in traditional dental restorations manufacturing is an emerging application. The objective of this study was to identify the different structural and morphological characteristics of Co-Cr dental alloys processed by alternative manufacturing techniques in order to understand the influence of microstructure on restorations properties and their clinical behavior. Metallic specimens made of Co-Cr dental alloys were prepared using traditional casting (CST), and computerized milling (MIL), selective laser sintering (SLS) and selective laser melting (SLM). The structural information of the samples was obtained by X-ray diffraction, the morphology and the topography of the samples were investigated by Scanning Electron Microscopy and Atomic Force Microscope. Given that the microstructure was significantly different, further differences in the clinical behavior of prosthetic restorations manufactured using additive techniques are anticipated.

1. Introduction

Cobalt-chromium (Co-Cr) alloys are wide used in the technology of fixed dental prosthesis because of their excellent mechanical properties and low production costs compared to high noble dental alloys. Continuous advancements in metallurgical technology have expanded the range of methods by which these restorations can be manufactured [1]. Main alternatives for the conventional melting-casting technology are digitalized processing technologies. The recent introduction of these techniques in the dental field has marked a turning point in the prosthesis production methodology, from the traditional hand made to the innovative automated approach. The integration of new technologies in traditional dental restorations manufacturing is an emerging application [2].
Digitized technologies can be classified as based on subtractive manufacturing (SM), such as the milling of pre-manufactured materials assisted by computer-aided design/computer-aided manufacturing (CAD/CAM) systems [3-8], or on additive manufacturing (AM), such as the more recently developed selective laser sintering (SLS) and selective laser melting (SLM) techniques [9-13].

Although CAD/CAM has long been directly associated with the milling procedure in dental literature, it should be mentioned that SLS and SLM are also classified as CAD/CAM technologies.
Processing of Co-Cr dental alloys can markedly affect the mechanical properties of the resulting components; therefore it should be carefully controlled to obtain the desired quality [14,15]. Although casting remains the traditional processes used to manufacture CoCr alloy restorations [14], nowadays additive manufacturing (AM) is being accepted as the novel candidate for the fabrication of customized dental prostheses[16-18]. AM has attracted much attention over the past ten years due to its immanent advantages, such as design freedom and short working times. AM techniques have already been known for more than 20 years but were at first limited to the rapid manufacturing of porous structures and prototypes [19].
Given the large differences in the manufacturing process between casting, which uses the complete melting and overheating of casting materials, the milling of a prefabricated metal block and additive manufacturing using a fine metallic powder, large differences in microstructural characteristics are anticipated [3]. In terms of chemical composition, there is little difference between Co-Cr alloys used for different manufacturing technologies. Typical Co–Cr alloys contain between 55-63 wt% cobalt and 25-28 wt% chromium [20]

2. Objective
The objective of this study was to identify the different structural and morphological characteristics of Co-Cr dental alloys processed by alternative manufacturing techniques in order to understand the influence of microstructure on restorations properties and their clinical behavior.

3. Materials and methods
Metallic specimens made of Co-Cr dental alloys were prepared using traditional casting (CST), and computerized milling (MIL), selective laser sintering (SLS) and selective laser melting (SLM).
For the experimental analyses round plates 20 mm diameter and 2 mm thick were fabricated using different technologies.
The cast alloy was manufactured by conventional lost wax technique with a phosphate-bonded investment. The mold was cast with Wirobond® SG alloy (Bego, Bremen, Germany) at 900°C using a vacuum pressure casting machines Nautilus (Bego, Bremen, Germany). The mold was left to cool down to room temperature and the specimens were then divested, sandblasted with alumina particles (200 μm) and finished with rotative instruments, burs suitable for Co-Cr alloys. Type 4 alloy Wirobond® SG fulfils all criteria of the ISO 22674 and ISO 9693-1 Standards. The composition is Co 63.8; Cr 24.8; W 5.3; Mo 5.1; other constituents Si, and Fe. Wirobond® SG allows normal cooling and provides effective, economic processing in the laboratory.
A prefabricated block of a commercial Co-Cr dental alloy CopraBond K (White Peaks Dental Systems GmbH & Co. KG, Essen, Germany), a non precious blank on Co-Cr base, type 4, was milled to fabricate the sample using Datron D5 5-axis dental milling machine (Datron, Mühltal, Germany). A disc wax pattern was digitized and the specimens were cut to their final dimensions. The composition is: Co 61.0, Cr 28.0, W 8.5, Mn 0.25, Fe <0.5, Si 1.65, C <0.1. CopraBond K is a nickel- and beryllium free Co-Cr blank, specially designed for CAD/CAM applications. The material is very homogenous and lends itself to machining extremely well by milling.
The laser-sintered specimens were prepared from commercial Co-Cr powder ST2724G (SINT-TECH,) using a dental laser sintering device (PXS Dental System, Phenix Systems, Clermont-Ferrard, France). For scanning the high accuracy D700 scanner (3Shape, Copenhagen, Denmark) and for design Dental System™ CAD Software (3Shape, Copenhagen, Denmark) were used. Chemical composition of Co-Cr alloy (wt.%): Co 54.31; Cr 23.08; Mo 11.12; W 7.85; Si 1.67; Mn 1.67; Fe<0,1. Cobalt-Chromium ST2724G is a type 5 alloy, certified ISO 9001 and ISO 13485. The layering function for the powder material is a fundamental part of the laser sintering process. A mechanism patented by Phenix Systems allows you to adapt to the shape and average size of the grains of powder. This function can be

programmed and adjusted to suit any type of ceramic or metal powder with a granulation equal to or greater than a micron.
For SLM a Co-Cr bonding alloy for the manufacturing of removable and fixed restorations by SLM was used. The alloy is a type 4 alloy according to ISO 22674. Nominal values of alloy composition in mass percent: Co 59.0; Cr 25.0; W 9.5; Mo 3.5; Si max. 1; other constituents: C, Fe, Mn, N max. 1.5.
The specimens resulted after CAD/CAM technologies were not additional prepared.
The weights of the specimens were: 1.40 g for MIL; 1.35 g for SLS; 1.17 g for SLM and 1.06 for CST.
The structural information of the samples was obtained by X-ray diffraction on a XRD, X'pert Pro MPD X-ray diffractmeter (PANalytical B.V., Almelo, Netherlands) using Cu Kα radiation, at room temperature. The morphology and the topography of the samples were investigated by Scanning Electron Microscopy with Energy Dispersive X-Ray Analysis SEM / EDAX Model INSPECT S (FEI, Oregon, USA) and atomic force microscope AFM Model Nanosurf® EasyScan 2 (Nanosurf AG, Liestal, Switzerland).

4. Results and discussions
Depending on the nature and chemical composition of the samples the morphology is different. The microstructure of Co–Cr dental alloys depends also on the manufacturing technique.
The morphological aspect of the resulting powders was examined by SEM, as shown in Figure 1 a, b, c, d. The magnification used for SEM measurements was 6000 X.
Depending on the nature and chemical composition of the samples the morphology is different.

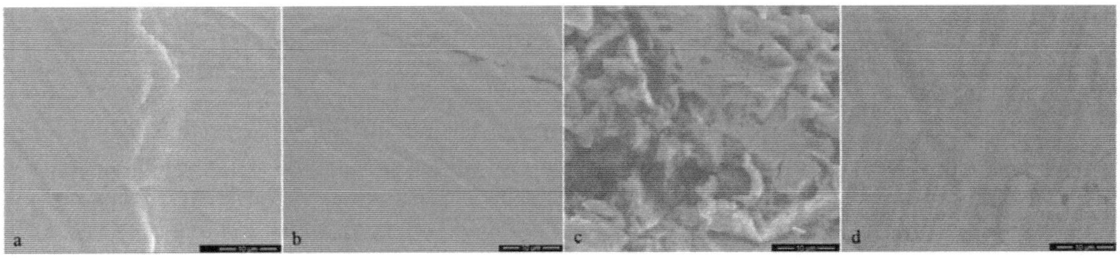

Figure 1. The SEM images of: a) MIL; b) SLS; c) SLM and d) CST.

In Figure 2 (a-d) the EDAX analyses of the samples are presented. The EDAX analysis certified initial molar ratios between material compositions.

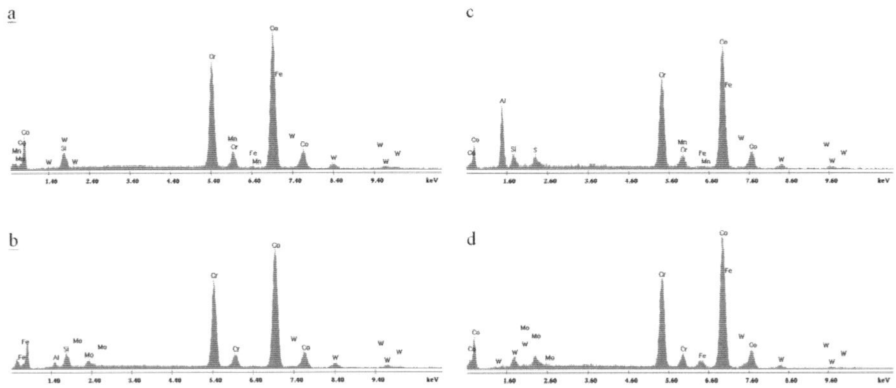

Figure 2. EDAX analysis of: a) MIL; b) SLS; c) SLM and d) CST.

In Figure 3 are shown the AFM images of the surface for each of the studied materials, using a scan size of 2µm x 2µm. The contact mode cantilever was used to measure the samples.

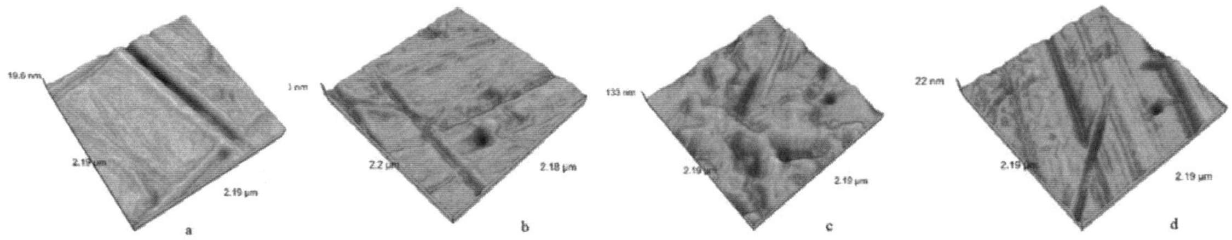

Figure 3. 3D AFM images of: a) MIL; b) SLS; c) SLM and d) CST.

The average surface roughness Sa and mean square root roughness Sq were calculated and for the samples and are given in Table 1.

Table 1. The surface roughnesses.

Samples	Sa (nm)	Sq (nm)
MIL	2.7	3.7
SLS	3.8	5
SLM	35	42
CST	7.4	8.8

There is a strong correlation between the aspect of the samples (SEM images) and the rugosity. The more uniform samples reflect in the lowest value of rugosity and the non uniform architectures present the highest value of roughness. The roughness is reflected in porous structures, which can be correlated also with the weight of the samples (in descending order MIL, SLS, SLM and CST).
These findings regarding microstructural properties may have clinical implications. SLM and SLS technologies have recently been introduced in the dental field, and a vast spectrum of factors should be tested and/or optimized to increase its efficacy in the production of metallic dental restorations [3].
In the steadily increasing market for 3D printing technologies, SLS and SLM solutions are leading procedures for metal based additive manufacturing technology. These extremely reliable and economical methods enable the manufacturing of highly complex geometry restorations, which cannot be milled in such a way. The material science analysis of cross sectional images confirms the outstanding quality of our metal powders. In the manufacturing process of the SLM technology the components achieve even better metallurgical characteristics than using the conventional casting process with the same material [21]. Taking into account the many benefits of AM technologies, and the findings regarding the studied structural and morphological characteristics of Co-Cr dental alloys processed by alternative manufacturing techniques, more research is required in order to understand the influence of microstructure on restoration properties, further veneering procedures of the frameworks, and last but not least the clinical behavior.

5. Conclusions

1. Even if there is little difference between the chemical compositions of the Co-Cr alloys used for different manufacturing technologies, the structural and morphological characteristics of Co-Cr dental alloys processed by alternative manufacturing techniques are different.
2. Milling of pre-manufactured blocks assisted by CAD CAM technologies is associated with the best structural and morphological characteristics of the processed materials.
3. The immanent advantages of additive technologies are accompanied by porosities resulted after the deposition of the fine metallic powder, especially for the SLM method. The chemical compositions resulting after AM actually differ widely from those indicated by manufacturers for the powders.
4. Given that the microstructure was significantly different, further differences in the clinical behavior of prosthetic restorations manufactured using additive techniques are anticipated.

Acknowledgments
This work was supported by a grant of the Romanian National Authority for Scientific Research and Innovation CNCS-UEFISCDI, project number PN-II-RU-TE-2014-4-0476.

References

[1] Li Kc, Prior Dj, Waddell Jn, Swain Mv. "Comparison of the microstructure and phase stability of as-cast, CAD/CAM and powder metallurgy manufactured Co-Cr dental alloys", Dent Mater. 31(12):306-15 (2015).
[2] Atzeni E, Salmi A. "Evaluation of additive manufacturing (AM) techniques for the production of metal–ceramic dental restorations", Journal of Manufacturing Processes 20:40-5 (2015).
[3] Al Jabbari Ys, Koutsoukis T, Barmpagadaki X, Zinelis S. "Metallurgical and interfacial characterization of PFM Co-Cr dental alloys fabricated via casting, milling or selective laser melting", Dent Mater;30(4): 79-88 (2014).
[4] Van Noort R. "The future of dental devices is digital". Dent Mater;28:3-12 (2012).
[5] Willer J, Rossbach A, Weber Hp. "Computer-assisted milling ofdental restorations using a new CAD/CAM data acquisitionsystem", J Prosthet Dent, 80:346-53 (1998).
[6] Miyazaki T, Hotta Y. "CAD/CAM systems available for thefabrication of crown and bridge restorations", Aust Dent J, 56(1):97-106 (2011).
[7] Witkowski S, Komine F, Gerds T. "Marginal accuracy of titanium copings fabricated by casting and CAD/CAM techniques", J Prosthet Dent, 96:47-52 (2006).
[8] Alt V, Hannig M, Wostmann B, Balkenhol M. "Fracturestrength of temporary fixed partial dentures: CAD/CAMversus directly fabricated restorations", Dent Mater 27:339–47 (2011).
[9] Quante K, Ludwig K, Kern M. "Marginal and internal fit ofmetal–ceramic crowns fabricated with a new laser melting technology", Dent Mater 24:1311–5 (2008).
[10] Traini T, Mangano C, Sammons Rl, Mangano F, Macchi A,Piattelli A. "Direct laser metal sintering as a new approach to fabrication of an isoelastic functionally graded material for manufacture of porous titanium dental implants", Dent Mater 24:1525-33 (2008).
[11] Akova T, Ucar Y, Tukay A, Balkaya Mc, Brantley Wa. "Comparison of the bond strength of laser-sintered and cast base metal dental alloys to porcelain", Dent Mater, 24:1400-4 (2008).
[12] Ucar Y, Akova T, Akyil Ms, Brantley Wa. "Internal fitevaluation of crowns prepared using a new dental crownfabrication technique: laser-sintered Co–Cr crowns". JProsthet Dent 102:253-9 (2009).
[13] Tian Xy, Gunster J, Melcher J, Li Dc, Heinrich Jg. "Process parameters analysis of direct laser sintering and posttreatment of porcelain components using Taguchi's method", J Eur Ceram Soc 29:1903-15 (2009).
[14] Monroya K, Delgadoa J, Ciurana J. "Study of the pore formation on CoCrMo alloys by selective laser melting manufacturing process", Proc Eng 63:361-9 (2013).
[15] Qian B, Saeidi K, Kvetková L, Lofaj F, Xiao C, Shen Z. "Defects-tolerant Co-Cr-Mo dental alloys prepared by selective laser melting", Dent Mater 31(12):1435-44 (2015).
[16] Bertol Ls, Júnior Wk, Da Silva Fp, Kopp Ca. "Technical Report Medical design: direct metal laser sintering of Ti–6Al–4V", Mater Des 31:3982-8 (2010).
[17] Senthilkumaran K, Pandey Pm, Rao Pvm. "Influence of building strategies on the accuracy of parts in selective laser sintering", Mater Des 30:2946-3295 (2009).
[18] Abd-Elghany K, Bourell Dl. "Property evaluation of 304 stainless steel fabricated by selective laser melting", Rapid Prototyp J 18(5):420-8 (2012).
[19] Herzog D, Seyda V, Wycisk E, Emmelmann C. "Additive manufacturing of metals", Acta Materialia 117:371-92 (2016).
[20] Okamoto H. Co–Cr (cobalt–chromium). J Phase Equilib, 24:377-8 (2003).
[21] https://slm-solutions.com/sites/default/files/attachment/page/2016/01/150305_slm_dental_englisch.pdf

Kinetic of carbonic anhydrase immobilized onto amberlite xad 7 and it application in sequestration of co$_2$ into caco$_3$ precipitate

Siti Noor Hazirah Mohd Rafidi, Siti Nadia Abdullah and Fazlena Hamzah*

Biocatalysis & Biobased Material Research Group, Green Technology and Sustainable Development Research Community, Faculty of Chemical Engineering, Universiti Teknologi Mara UITM, 40450 Shah Alam, Selangor, MALAYSIA

*fazlena@salam.uitm.edu.my

Abstract. Carbonic Anhydrase (CA) was immobilized onto Amberlite XAD 7 and was used in carbon dioxide sequestration purposes. The catalytic activities for free and immobilized CA were estimated by using para-Nitrophenyl Acetate (p-NPA) as the substrate in Tris-buffer containing 10% of acetonitrile. Lineweaver-Burk plot was employed to estimate the Michaelis-Menten kinetic parameters for both free and immobilized CA. K_m value of free and immobilized CA were 2.92 mM and 5.7 mM respectively. Meanwhile the V_{max} value of free and immobilized CA were 5.95 μmoles/min/ml and 2.67 μmoles/min/ml respectively. The kinetic value obtained in the present study shown that the immobilized CA has high affinity for its substrate. On the other hand, activity and stability study at various pH and temperature indicates that the optimum pH for free CA was found to be at pH 9 while for immobilized CA was at pH 10. For optimum temperature, a free CA was performed optimally at temperature 25°C and immobilized CA was working effectively at temperature 50°C. The immobilized CA onto Amberlite was tested in the CO_2 sequestration process and the formation of the white $CaCO_3$ precipitate was observed during the process. $CaCO_3$ powder obtained during the process was validated with the XRD analysis. The finding indicated that immobilized CA onto Amberlite XAD7 retained it enzymatic activity and stability and thus perform well in the CO_2 sequestration which gave the white $CaCO_3$ precipitate.

1. Introduction

Nowadays, global warming becomes one of the major problems to the earth. One of the major contributors of the greenhouse gases is carbon dioxide. These happen due to the rising of anthropogenic CO_2 that has been emitted by the industrial sector such as ammonia production, natural gas processing, cement manufacturer and many others. According to Calleja et al, [1] the emission of carbon dioxide from fossil fuel plants is the main contributor to the problem arises. In 2007, the Intergovernmental Panel on Climate Change (IPCC) has established Carbon Capture and Storage (CCS) as an option for stabilization of greenhouse gas concentration while allowing for continuation of using fossil fuel [2]. CCS concept is based on the reduction of CO_2 emissions to the atmosphere produce by industrial processes. There are three major stages in CCS which are; i) capture of CO_2 gas in atmosphere, ii) transport of CO_2 storage place and iii) storage of captured CO_2. There are three major techniques of CCS which categorized as post-combustion, oxy-combustion and pre-combustion.

Technology of CCS involves several processes such as physisoprtion/chemisorption, membrane separation/molecular sieve, carbamation, amine physical absorption, amine dry scrubbing, mineral carbonation. Unfortunately, the existing technology has several drawbacks such as in amine technology, there are limitations on insufficient carbon dioxide storage, equipment corrosion and many others [3].

Moreover, investigation on expenses of conventional CCS technologies has found that most of the system is costly. The estimated cost of the process shown in Figure 1 is around \$41-\$72 per ton of CO_2 sequestrated. This cost can be reduced if the captured CO_2 can directly convert into valuable product without bothering any further compression, transportation and injection processes. Thus, a study on enzymatic carbon capture becomes one of the alternatives to tackle this limitation. Muradov (2012) has suggested the reaction between carbonic anhydrase and CO_2 to form bicarbonate product which is more practical and valuable in the market. Carbonic anhydrase (CA) catalyzes the conversion of carbon dioxide and water into bicarbonate via a hydration mechanism. Biocatalyst offers a good opportunity due to its high level of catalytic efficiency and a high degree of selectivity. Unfortunately, there are some limitations with biocatalyst such as high cost, limited stability, difficulty in handling, and so on. Thus, to overcome the problems, enzyme needs to be immobilized into the support material to enhance its usability and reaction [5], [6]. Many techniques of enzyme immobilization have been developed wherein adsorption on solid support material is the easiest and the conventional method. In this paper, Amberlite XAD-7 is being extensively studied for enzyme immobilization by virtue of its non-toxic.

Amberlite XAD 7 is a porous resin that is used to immobilize enzymes via simple adsorption. To help the attachment of the enzyme into the Amberlite XAD 7, gluteraldehyde cross-linking was used to bind the enzyme via covalent bond. To the best of our knowledge, detailed studies on immobilization of carbonic anhydrase on Amberlite XAD-7 have not been reported so far. In this paper, CA was immobilized onto Amberlite XAD7 and detailed kinetic parameters like pH, temperature of free and immobilized was investigated. Characterization of immobilized enzyme was done using SEM and FTIR. While calcium carbonate precipitate was characterized using XRD for the validation process.

2. Methodology

2.1. pH and Thermal stability

The immobilized enzyme was prepared according to the Wanjuri et al [10] and 0.2g immobilized enzyme was weighed for thermal and pH stability study. 13.5 ml buffer pH 7 and 1.5 ml p-NPA solution were added into the beaker. The experiment for the pH stability was carried out for 2 hours and 15 minute interval for absorbance analysis at 400 nm. The similar step was repeated with pH 6, 8, 9 and 10. For thermal stability, the temperature of the reaction was set at six different temperatures, which are 30°C, 40°C, 50°C, 60°C and 70°C. Similar with pH stability, 15 ml of reaction volume was used with 13.5 ml of pH buffer, and 1.5 ml p-NPA solution as well as 0.2 g of immobilized enzyme was used. The absorbance was taken for every 15 minutes of interval for 2 hour duration. The absorbance reading was taken at 400 nm wavelength.

2.2. Kinetic study

For immobilized CA the buffer volume reaction used was 15 ml with reaction temperature set at 25°C and pH 7.5. p-NPA solution with different concentration from 2.5 mM, 2.0 mM, 1.5 mM, 1.0 mM and 0.5 mM was prepared. The absorbance reading was taken for 2 hour with 15 minute interval. The absorbance reading was taken at 400 nm.

2.3. CaCO3 Precipitation

The main experiment was started with the preparation of 200 ml of $Ca(OH)_2$ solution with concentration of 30 mM. During the precipitation reaction, the reading of pH was taken for every 15 seconds with constant flow rate and rpm (1L/min and 200 rpm). The reaction was stop when the pH

IOP Conf. Series: Journal of Physics: Conf. Series **885** (2017) 012006 doi:10.1088/1742-6596/885/1/012006

achieved 7. The carbonation reaction was tested for variation of temperature form 30°C, 50°C, 70°C and 85°C. For the characterization of the precipitation, the solution was filtered and dried overnight. The dried precipitation was examined using XRD.

3. Results and discussion

3.1. pH and temperature stability

Knowing the optimum pH is an important factor as it is a crucial parameter for an enzyme to work at it best performance. Besides, pH also plays a part in altering enzyme activity in aqueous solution. Figure 1(a) shows the effect of pH of free and immobilized CA to its activity. For free enzyme, the optimum pH of CA was obtained at pH 9 while for immobilized CA, the optimum pH was at 10. From the study, the optimum pH was changed from free cell to the immobilized cell. The optimum pH of many enzymes is shifted to higher pH if carrier used is anionic and towards low pH if it is cationic. In this study, the carrier used for immobilization were anionic thus it changes the degree of ionization of amino acid residues at the active site which lead to change in electrostatic potential and shift the pH towards alkaline [8] [7]. Figure 1(b) present the thermal stability profile of both free and immobilized CA. The results clearly showed that the activity of free enzyme was optimum at temperature 25°C and gradually decreases as the temperatures approaches 70°C. Meanwhile, the activity of immobilized CA was increased and optimum at temperature 50°C before the activity was dropped when temperature 60°C and 70°C was used in the system. From results, it was observed that the loss of activity for immobilized enzyme was lower compared to the free enzyme. This phenomena happen due to the present of carrier(Amberlite) that has protecting the enzyme when the temperatures was increased and eliminated the potential of enzyme deactivation to occur. Immobilization is the reason to the conformational flexibility of the enzyme thus lead to the increase in enzyme rigidity which commonly directly increases the enzyme stability towards a high temperatures[9] [8]. Moreover, it is often observed that immobilized enzyme has higher thermal stability than free enzyme due to the conformational flexibility in immobilized enzyme [10] [9].

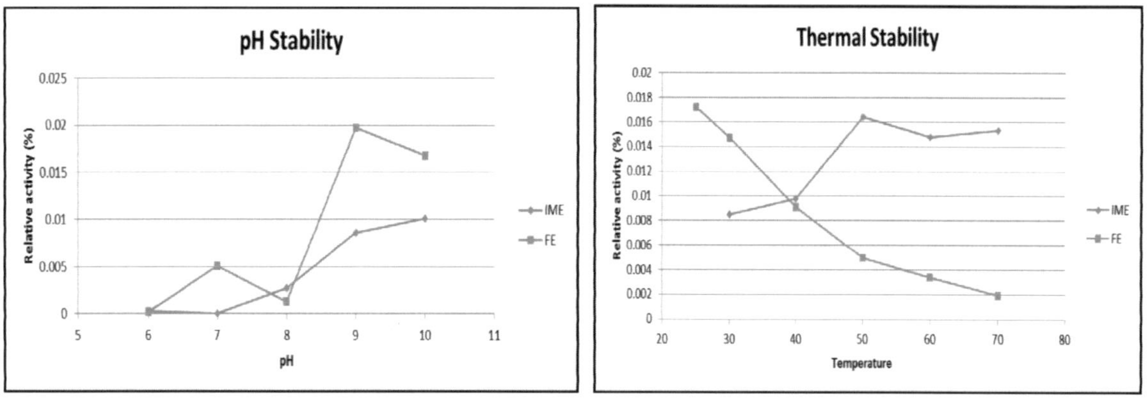

(a) (b)

Figure 1. (a) pH stability profile for free CA and immobilized CA (b) Temperature stability profile for free CA and immobilized CA

3.2. Kinetic study

The kinetic constants (K_m and V_{max}) for free and immobilized enzyme were determined by using Lineweaver-Burk plots. The values of K_m and V_{max} of free and immobilized CA were calculated from the intercepts of x and y axes of the Lineweaver-Burk plot respectively. The values of K_m calculated for both free and immobilized CA were 2.919 mM and 5.711 mM respectively while V_{max} for free and immobilized CA were 5.952 μmoles/min/ml and 2.666 μmoles/min/ml. Note that the value of K_m for free is always lower than the immobilized enzyme. The experiment conducted by S. Wanjari et al [7] [10], indicated that the value of K_m for free enzyme is much lower compared to the K_m value of immobilized enzyme. K_m is defined as the substrate concentration at which the reaction rate is half of its maximum value. Therefore, in other words, smaller value of K_m indicates that the enzyme has achieves its maximum catalytic efficiency at low substrate concentrations. Thus, the smaller the K_m, the more efficient the enzyme works [11]. From the results, free CA has lower value of K_m as compared to immobilize CA. From the statement, it can be concluded that free CA has higher catalytic efficiency compared to immobilize CA. There are several factors that contribute to the change of kinetic parameter of immobilize enzyme such as conformational changes and steric effect [12].

3.3. $CaCO_3$ precipitation

In order to test the capability of immobilized enzyme in carbonation process, studies were carried out using a buffering process. In the carbonation reaction, it was observed that immobilized CA react with CO_2 to gave white precipitate. The XRD analysis conducted show that the white precipitate formed was $CaCO_3$. The XRD pattern of precipitate products indicated that all three calcium carbonate precipitated are indeed calcite. The regions of peaks show that atomic arrangements are the same for the three precipitated calcium carbonate. This result indicates that the transformation of CO_2 to $CaCO_3$ was possible by the involvement of immobilized CA which catalyzes the formation of bicarbonate ion, HCO3-. Similar mechanism of carbonation by carbonic anhydrase was reported by Favre et al [13].

4. Conclusion

The Carbonic anhydrase (CA) enzyme immobilized onto Amberlite XAD 7 has shown a potential to be used in hydration of CO_2 based on its activity and stability on various pH and temperature. The study indicated that the optimum temperature for immobilized CA was at 50°C and the optimum pH was at 9. The value obtained was higher compared to stability of free CA. the The catalytic activity of immobilized enzyme calculated using Michaelis Menten was 5.7mM and 2.67 μmoles/min/ml for K_m and V_m value respectively. The kinetic value obtained in the present study shown the capability of the immobilized CA to be used in the reaction process. The immobilized CA was used in CO_2 hydration process and the results show that $CaCO_3$ precipitation was formed as a product of the reaction. The formation of $CaCO_3$ validated the present of CA onto Amberlite and its capability to be used as a catalyst in CO_2 sequestration process.

Acknowledgments

The present study was made possible through funding from Universiti Teknologi MARA - 600-IRMI/GIP 5/3 (0041/2016) and continuous support from Research Management Institute (RMI) of Universiti Teknologi MARA is gratefully acknowledged. Authors also like to show gratitude to the facilities and constant encouragement from Faculty of Chemical Engineering, Universiti Teknologi MARA.

References

[1] E. Torralba-Calleja, J. Skinner, and D. Gutiérrez-Tauste, "CO2 Capture in Ionic Liquids: A Review of Solubilities and Experimental Methods," J. Chem., vol. 2013, pp. 1–16, 2013.
[2] T. Barker, "Climate Change 2007 : An Assessment of the Intergovernmental Panel on Climate Change," Change, vol. 446, no. November, pp. 12–17, 2007.
[3] O. F. Dawodu and A. Meisen, "Degradation of alkanolamine blends by carbon dioxide," Can. J. Chem. Eng., vol. 74, no. 6, pp. 960–966, 1996.
[4] N. Z. Muradov, "Fossil fuel decarbonization: in the quest for clean and lasting fossil energy.," in Carbon-Neutral Fuels Energy Carriers, 2012, pp. 667–786.
[5] S. Hosseinkhani and M. Nemat-Gorgani, "Partial unfolding of carbonic anhydrase provides a method for its immobilization on hydrophobic adsorbents and protects it against irreversible

thermoinactivation," Enzyme Microb. Technol., vol. 33, no. 2–3, pp. 179–184, 2003.
[6] F. Azari, S. Hosseinkhani, and M. Nemat-Gorgani, "Use of reversible denaturation for adsorptive immobilization of urease," Appl Biochem Biotechnol, vol. 94, no. 3, pp. 265–277, 2001.
[7] S. Wanjari, C. Prabhu, T. Satyanarayana, A. Vinu, and S. Rayalu, "Immobilization of carbonic anhydrase on mesoporous aluminosilicate for carbonation reaction," Microporous Mesoporous Mater., vol. 160, no. 4, pp. 151–158, 2012.
[8] S. Kumar, R. K. Yadav, and S. Negi, "A comparative study of immobilized lipase produced from Penicillium chrysogenum SNP5 on two different anionic carriers for its pH and thermostability," Indian J. Biotechnol., vol. 13, no. 3, pp. 301–305, 2014.
[9] M. A. Abdelnaby, "Immobilization of Aspergillus-Niger Nrc-107 Xylanase and Beta-Xylosidase, and Properties of the Immobilized Enzymes," Appl. Biochem. Biotechnol., vol. 38, no. 1–2, pp. 69–81, 1993.
[10] Ş. Akkuş Çetinus and H. Nursevin Öztop, "Immobilization of catalase into chemically crosslinked chitosan beads," Enzyme Microb. Technol., vol. 32, no. 7, pp. 889–894, 2003.
[11] W. Cleland, "Enzyme kinetics 1," Wiley Encycl. Chem. Biol., vol. 6, no. 2, pp. 1–11, 1967.
[12] J. F. Kennedy, Immobilized Enzyme Principles (Applied Biochemistry and Bioengineering, Volume 1), vol. 67, no. 2. Elsevier, 1978.
[13] N. Favre, M. L. Christ, and A. C. Pierre, "Biocatalytic capture of CO_2 with carbonic anhydrase and its transformation to solid carbonate," J. Mol. Catal. B Enzym., vol. 60, no. 3–4, pp. 163–170, 2009.

IOP Conf. Series: Journal of Physics: Conf. Series **885** (2017) 012007 doi:10.1088/1742-6596/885/1/012007

Determining the size of nanoparticles in the example of magnetic iron oxide core-shell systems

Maciej JARZĘBSKI[1], Mikołaj KOŚCIŃSKI[2], Tomasz BIAŁOPIOTROWICZ[1]

[1] Off-Campus Faculty of Law and Social Sciences in Stalowa Wola Catholic University of Lublin, Kwiatkowskiego 3A, 37-450 Stalowa Wola, Poland
[2]Department of Physics, Poznan Life Science University, Wojska Polskiego 38/42, Poznań, Poland

maciej.jarzebski@o2.pl, jarzebski@kul.lublin.pl

Abstract. The size of nanoparticles is one of the most important factors for their possible applications. Various techniques for the nanoparticle size characterization are available. In this paper selected techniques will be considered base on the prepared core-shell magnetite nanoparticles. Magnetite is one of the most investigated and developed magnetic material. It shows interesting magnetic properties which can be used for biomedical applications, such as drug delivery, hypothermia and also as a contrast agent. To reduce the toxic effects of Fe3O4, magnetic core was covered by dextran and gelatin. Moreover, the shell was doped by fluorescent dye for confocal microscopy investigation. The main investigation focused on the methods for particles size determination of modified magnetite nanoparticles prepared with different techniques. The size distribution were obtained by nanoparticle tracking analysis, dynamic light scattering and transmission electron microscopy. Furthermore, fluorescent correlation spectroscopy (FCS) and confocal microscopy were used to compare the results for particle size determination of core-shell systems.

1. Introduction

Development of nanotechnology opens new possibilities in the design, formulation and manufacture of structures and materials with desirable physicochemical properties. Nanomaterials are considered according to their unique behavior as an innovative solutions in the fields of medical applications, electronic devices manufacturing, sensors, energy storage, daily use devices and so on [1-2]. One of the most important factor which decides of the final application of NPs is the size (mostly denoted by diameter of the particle).

Magnetic iron nanoparticles have been investigated for several decades as possible transporters of a therapeutic substance. The concept is, the possibility to control the direction of flow of particles with magnetic properties, using the external electromagnetic field. The key role plays two iron oxide structures: maghemite γ -Fe_2O_3 and magnetite Fe_3O_4, which exhibit magnetic properties. MNP are usually considered for their use in magnetic separation, diagnostics, including MRI (magnetic resonance imaging) or hyperthermia and tissue engineering [3]. The bioavailability and possible application for biomedical systems [4] of MNP strongly depends on their size and structure. This is one of the major intent to use various techniques for a particle size characterization.

From the definition, one of the dimension of nanosystem should be in the range between 1 – 100nm. In practice the structures larger than 100nm but smaller than 1µm are named as a nanostructures, too. In this paper systems which the size above 100nm will be described also as submicronsized to emphasize and more clearly discuss the difference between the particle size determination techniques. As a representative nanostructures for this case study we chose the core-shell MNP. As the core magnetic iron oxide nanoparticles were prepared with the selected commonly used methods. Biopolymers, such as dextran and gelatin were used as a shell. Fluorescent agents, such as Rhodamine B isothiocyanate (RBITC) and Fluorescein isothiocyanate (FITC) were added for biopolymeric matrix. This brought the opportunities to investigate the fluorescent behavior of CS-MNP with confocal laser microscopy (CLSM) and fluorescence correlation spectroscopy (FCS). In this paper the main differences between selected particle size determination methods will be discussed.

2. Materials and methods

2.1. Preparation of the core-shell nanoparticles
Magnetic iron nanoparticles were prepared with minor changes, according to the methods described by Hyeon et al. [5], Park et al. [6] and B. Gaihre et al. [7]. Next MNP cores were attached by biopolymeric shells based on fluorescent dextran and gelatin [8-9].

2.2. Particle size determination techniques
Hydrodynamic diameters (dH) of the MNPs and CS-MNPs were obtained using Dynamic Light Scattering (DLS) and Nanoparticle Tracking Analysis (NTA). For the DLS and NTA investigations NanoSight NS500 with laser beam wavelength of 405 nm were used. Before the examinations the MNP and CS-MNP were suspended in water.
For structural investigations a Transmission Electron Microscope (TEM) - JEOL JEM-1011 microscope was used in UniHamburg.
Confocal Laser Scanning Microscopic studies were performed using confocal microscope (Zeiss Laser Scanning Microscope LSM780). For the measurements, the chamber coverglass (8 WELL from LAB-TEK) was used and or filled up slide channels (1 µ-Slide from Ibidi) were used. Moreover Fluorescence Correlation Spectroscopic (FCS) measurements were made at the same CLSM with special module ConfoCore3 and software for autocorrelation function (AF) determination.

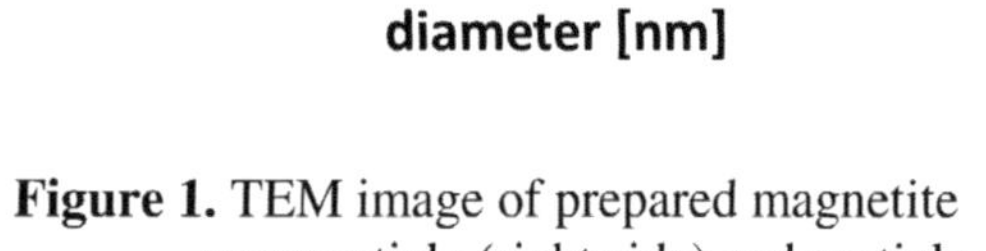

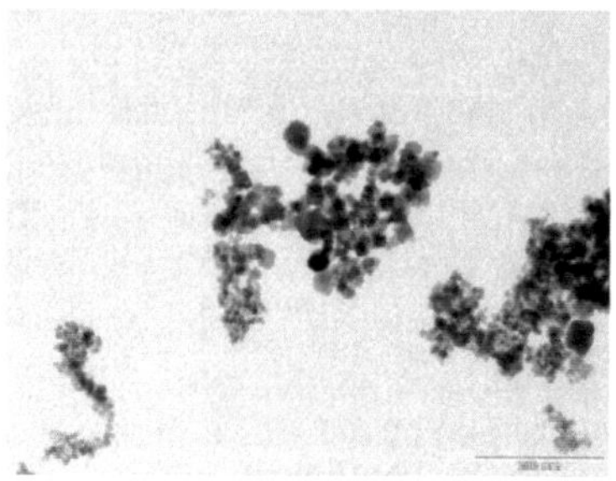

Figure 1. TEM image of prepared magnetite nanoparticle (right side) and particle size distribution from TEM (left side)

3. Results
Iron oxide nanoparticles were synthesized using 3 methods. The modification was conducted using dextran and gelatin as a shell for iron oxide particles. The fluorescent agent, which was a small volume of rhodamine or fluorescein labeled dextran or rhodamine labeled gelatin, was added to the biopolymer shell in order to enable confocal microscope investigations of the particles.

For this paper only selected and representative samples have been chosen of MNP – magnetic cores and MNP-CS –magnetic cores with fluorescent biopolymeric shell. TEM images (Figure 1 and 2a,c) of the prepared samples, confirmed that the synthesized magnetic nanoparticles (MNP) with a diameter in the range from 5nm to 50nm. Smaller particles exhibited lower polydispersity than those with the larger diameter (see Figure 2a). The cores (MNP) were next modified by covering them with labeled and unlabeled dextran and gelatin. Previously synthesized cores were mixed with the 5% solution of dextran or gelatin and vigorously stirred for several hours. Then the mixture was centrifuged several times and dialyzed overnight. The TEM images confirmed the successful preparation of the core-shell system for hydrophilic MNP.

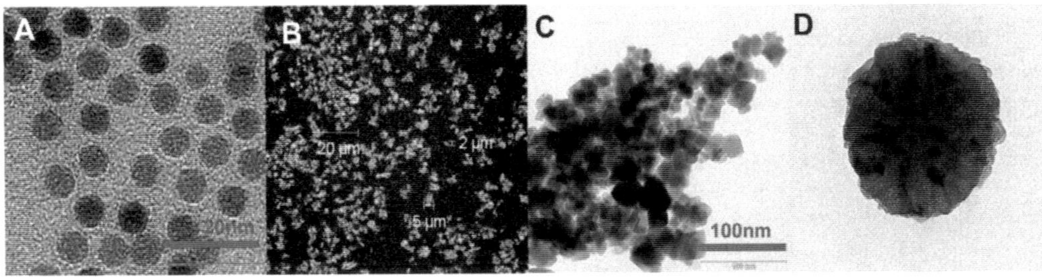

Figure 2. TEM and CLSM images of core and core-shell MNP nanoparticles: A. MNP-core with low polydispersity (TEM), B. MNP-CS with high polydispersity (CLSM), C. MNP-core with high polydispersity (TEM), D. MNP-CS with high polydispersity (TEM)

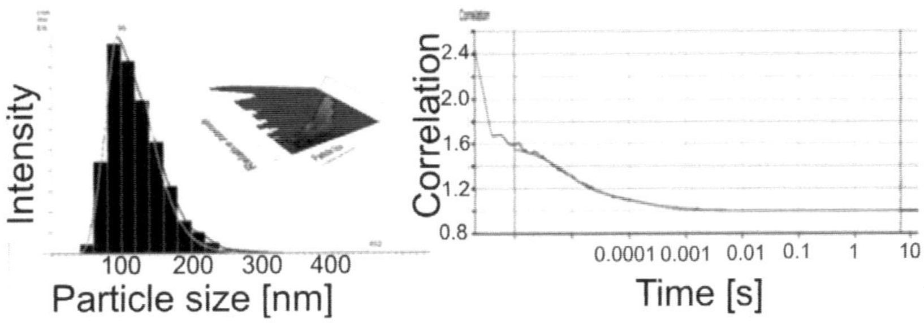

Figure 3. Particle size determination of MNP-CS, on the left: NTA anlaysis (2D plot: black bars – distribution, yellow line – main peak and 3D plot: relative intensity versus particle size) and DLS (pink line); on the right FCS measurement of core-shell systems.

Detailed TEM imaging shows that the shape and possibility of agglomeration of a particle depends on the synthesis method. The most regular spherical shapes were obtained by the methods described in [5-6]. The highest polydispersity was determined after the synthesis from iron fluorides (see Figure 2) and after core-shell structures preparation with biopolymeric dextran and gelatin shell (see Figure 2B and 2D).

To confirm the particle size, hydrodynamic diameters were obtained with two techniques: Nanoparticle Tracking Analysis (NTA) and Dynamic Light Scattering (DLS). Even these two methods use the same Stokes-Einstein relation between the diffusion of the particles, hydrodynamic diameter and viscosity, the obtained results of dH are different [10]. Representative results are present in Table 1 and Figure 3(left). As it is presented, the hydrodynamic diameters are larger than determined from NTA. It is caused by the fact that the diffusion coefficient determined in NTA is independent on the molar concentration of the investigated particles (for more physical and mathematical aspects see [10-12]). Comparing TEM and NTA results it is proved that some of the samples exhibit high polydispersity.

Figure 2 (right) shows the autocorrelation function (AF) of MNP-CS which is determined with FCS technique. From the calculations after fitting the FCS results to the mathematical model, the final hydrodynamic diameter was around few nanometers. This results are outstanding from the TEM, NTA and DLS investigations. Nevertheless, in FCS technique only fluorescent part (particles) have been "seen". That indicates, that the size of the particles of fluorescent biopolymeric shells was obtained. Moreover, it is suggested that in the investigated sample the signal from core-shell MNP systems is too weak. Furthermore, it confirms that not whole the MNP cores where covered by the fluorescent shell and a lot of labelled biopolymer is in the sample.

Table 1. Representative results from NTA and DLS measurements of MNP and MNP-CS

Sample description		d_{DLSNS} [nm]	$d_{NTAmean}$ [nm]	$d_{NTAmode}$ [nm]	**SD** [nm]
core	shell				
MNP04	---	290	208	154	74
MNP02	dextran FITC M=6kD	99	122	95	40

4. Discussion

Particle size determination is one of the crucial step for their future applications. Nowadays a lot of techniques are available for nanoparticle characterization. Nevertheless, during the analysis of the same sample, such as in this study, some differences of the value of the particle diameters occur. Bell et al. [13] have been prepared critical evaluation of the emerging techniques for submicrometer particle sizing, using as a model silica particles. In this paper, additional technique like FCS was suggested for nanoparticle and submicron sized particles characterization. Moreover, the investigations present the results of fluorescent systems.

In general no best techniques have been developed so far and the results using different investigation methods often give different values of the particles size. The deep size and structure analysis of the nanoparticles is possible by the TEM. The size, shape and the particle size distribution can be confirmed. On the other hand, only selected area of the measured grid is investigated, which caused that lower number of the particles population is characterized. Moreover, usually for TEM the sample is dried or freeze which may cause the structure changes in some systems (i. e. hydrocolloids or polymers).

The highest concentration (the largest population) of the particles can be used in DLS (according the described techniques). With DLS, there is no limit for the solution in which particles are suspended. Unfortunately, DLS is sensitive for the dust and the impact of larger structure – in some cases, signal from the small particles can be screened by the larger particles. NTA shows better the relation between the population of two or more different particle sizes than DLS and single particle tracking is possible. With NTA fluorescent behavior of the investigated system can also be confirmed (depending on the laser wave length). For online investigations of the interaction in the system, confocal microscopy is recommended. It is usually used for the micron structures analysis and brings information (without size analysis) where the nanostructures accumulate or agglomerate creating large structures. CLSM plays a key role for biological, cell and food system investigations [14].

Typical confocal microscope with dedicated detector (in Zeiss LSM 780 - ConfoCore3 module) can be successfully used as a tool for fluorescence correlation spectroscopy (FCS) [15]. Our previously described investigations proved and experimentally validated with a simple procedure, allowing accurate FCS measurements for submicron sized particles characterization [16-17]. In the present study, results bring an information about the MNP-CS particle size and behavior.

5. Conclusions

Particle size, especially in the nanometer range, plays a key role for the final applications of newly designed material. Various techniques for nanoparticles characterization are available, but it is recommended to use more than one. Moreover, FCS might be successful applied as fluorescent particles size determination and possible contamination of fluorescent agents in comparison of TEM, SEM and DLS techniques. Knowing the opportunities and weaknesses of presented size determination technique it is possible to choose the optimal solution dedicated for measured system.

Acknowledgment
The investigations were supported from the ESMI project, which is supported by the European Commission through FP7 Project title: ESMI, European Soft Matter Infrastructure. Grant agreement number: 262348References

[1] R. Ravichandran "Nanotechnology-Based Drug Delivery Systems," Nanotechnology-Based Drug Delivery Systems", 5, (2009), 17-33
[2] R. Bogue "Nanosensors: a review of recent progress," Sensor Review, 28, 1, (2008), 12 - 17
[3] M. A. Phillips, M. L. Gran, N. A. Peppas, „Targeted nanodelivery of drugs and diagnostics," Nano Today, tom 5, p. 143—159, 2010.
[4] A. Mishra " Iron Oxide Nanoparticles for Biomedical Applications," International Journal of Mechanical Engineering and Robotics Research, 3, 2, (2014), 410-414
[5] T. Hyeon, S. S. Lee, J. Park, Y. Chung, H. Bin "Synthesis of Highly Crystalline and Monodisperse Maghemite Nanocrystallites without a Size-Selection Process," Journal of the American Chemical Society, 123, 51, (2001) 12798-12801
[6] J. N. Park, K. An, Y. S. Hwang, J.-G. Park, H.-J. Noh, J.-Y. Kim, J.-H. Park, N.-M. Hwang, T. Hyeon "Ultra-large-scale syntheses of monodisperse nanocrystals," Nature Materials, 3, (20014) 891-895
[7] B. Gaihre, M. S. Khil, D. R. Lee, H. Y. Kim "Gelatin-coated magnetic iron oxide nanoparticles as carrier system: Drug loading and in vitro drug release study," International Journal of Pharmaceutics, 365, (2009), 180–189
[8] B. Gaihre , M. S. Khil , H. K. Kang, H. Y. Kim "Bioactivity of gelatin coated magnetic iron oxide nanoparticles: in vitro evaluation," Journal of Materials Science: Materials in Medicine, 20, (2009), 573–581
[9] B. Gaihre, S. Aryal, N. A. M. Barakat, H. Y. Kim "Gelatin stabilized iron oxide nanoparticles as a three dimensional template for the hydroxyapatite crystal nucleation and growth," Materials Science and Engineering C, 28, (2008), 1297–1303
[10] T. Śliwa, M. Jarzębski, K. Szutkowski "Nanoparticle tracking analysis of latex standardized beads," Current Topics in Biophysics, 37, (2014), 49-53
[11] B. J. Berne, R. Pecora, Dynamic Light Scattering: with Applications to Chemistry, Biology, and physics, New York: Wiley, 1976.
[12] T. Śliwa, M. Jarzębski "Dynamic light scattering investigation of PNIPAM-co-MAA microgel solution" Current Topics in Biopysics, 37, (2014) 29-33
[13] N. C. Bell, C. Minelli, J. Tompkins, M. M. Stevens, A. G. Shard "Emerging Techniques for Submicrometer Particle Sizing Applied to Stöber Silica," Langmuir, 28, (2012) 10860−10872
[14] M. Jarzebski, B. Bellich, T. Bialopiotrowicz, T. Sliwa, J. Koscinski, A. Cesaro "Particle tracking analysis in food and hydrocolloids investigations," Food Hydrocolloid, 68, (2017), 90-101
[15] E. Banachowicz, A. Patkowski, G. Meier, K. Klamecka, J. Gapiński "Successful FCS Experiment in Nonstandard Conditions" Langmuir, 30, (2014), 8945−8955
[16] J. Gapinski, M. Jarzebski, J. Buitenhuis, T. Deptula, J. Mazuryk, A. Patkowski, "Structure and dimensions of core-shell nanoparticles comparable to the confocal volume studied by means of FCS", Langmuir, 32, 10, (2016) 2482-2491
[17] T. Deptula, J. Buitenhuis, M. Jarzebski, A. Patkowski, J. Gapinski "Size of Submicrometer Particles Measured by FCS: Correction of the Confocal Volume," Langmuir, 31, 24, (2015) 6681-6687

IOP Conf. Series: Journal of Physics: Conf. Series **885** (2017) 012008 doi:10.1088/1742-6596/885/1/012008

Effect of concentration of Curcuma longa L. on chitosan–starch based edible coating

N. M. Yusof[1], J. Jai[1], F. Hamzah[1], A. Yahya[1], S. Pinijsuwan[2]

[1]Faculty of Chemical Engineering, Universiti Teknologi MARA, Malaysia, 40450 Shah Alam, Selangor, Malaysia
[2]School of Agro Industry, Mae Fah Luang University, 57100 Chiang Rai, Thailand.

noorsuhana@salam.uitm.edu.my

Abstract. The ability of chitosan-starch based coating to extend shelf life of strawberry were studied. The main objectives of this paper is to study the effects of different concentrations (20, 15, 10 and 5 μL) of *Curcuma longa L.* (CUR) essential oil into chitosan-based edible coating on surface tension in order to increase the effectiveness of the coating. CUR or turmeric is one of the commercially planted herbs in Malaysia for its phytochemical benefits. Application of edible coating using dipping technique has been analysed and evaluated for their effectiveness in extending shelf life of fruits. Surface tension was analysed to investigate the adhesion properties. The best CUR concentration was 15 μL with the optimum surface tension was found to be 31.92 dynes/cm.

1. Introduction

Strawberry is a highly nutritious, juicy and delicious fruit but very perishable with short shelf life due to fungal decay. It is one of the most refreshing, popular and healthy treats indeed. Therefore, it is important to satisfy all demands by developing better methods in extending shelf life of strawberry without any preservatives so that they are all in good quality even after several days. Technologies for food packaging and coating using biopolymers contribute to an interesting alternative in reducing the application of chemical preservatives. Aider (2010) identified that among most polymers, chitosan is one of substance that is extremely reliable as additive for bio-based active film preparation. Thus, edible films and coatings from chitosan seems to be highly promising to extend post-harvest shelf life of strawberry fruit. Edible films and coatings are thin films applied on food products that play an important role to maintain the quality and can be eaten as part of the products (Falguera et al. 2011). It has been used to protect the fruits from any damages include physical, chemical and mechanical as well as microbiological activities. On top of that, edible films and coatings are also applied to many products to control moisture transfer, gas exchange or oxidation process to overcome negative effects on product quality such as browning, off-flavour development, texture breakdown and microorganisms attacks (Rojas-Graü, Soliva-Fortuny, and Martín-Belloso 2009).

Edible chitosan films and coatings can be incorporated with another active compound to ensure that they perform their level best. As reported by Azevedo et al. (2014), edible chitosan coatings enriched with essential oil have been shown a great result as an effective antimicrobial activity against some foodborne pathogens. The effect of antimicrobial activity can protect fruits from microbial spoilage so that the shelf life can be extended. Among all the perishable fruits, strawberry is considered to be the

IOP Conf. Series: Journal of Physics: Conf. Series **885** (2017) 012008 doi:10.1088/1742-6596/885/1/012008

most difficult to maintain the freshness and has therefore been selected to test the efficacy of edible coating (Sallato et al. 2007). Strawberry which is known as *Fragaria x Ananassa Duch* is highly consumed berries worldwide because of its flavour and preferred organoleptic properties (Šamec et al. 2016). It has unique shape, fascinating colour, aromatic and delightful taste.

2. Materials and method

A systematic methodology has been identified to facilitate the research objectives. The experimental works were done in three parts including the preparation of chitosan solution, starch solution and mixture of chitosan and starch solution.

2.1. Materials

Materials involved are cassava starch, *Curcuma longa L.* (CUR) was used as the essential oil from BF1 Sdn. Bhd., 0.1M sodium hydroxide, chitosan with 97% deacetylation, tween 80, acetic acid 99%, and glycerol.

2.2. Edible coating preparation

Chitosan-starch edible coating was developed based on the procedures by Vásconez et al.(2009) with some modifications.

2.2.1. Chitosan solution.

Chitosan with 1% (w/v) was gelatinized in an aqueous solution of 0.5% (w/v) acetic acid at room temperature. 0.1 % (w/v) of tween 80 was added as a surfactant to increase wettability (Santos et al. 2012). In order to ensure that the chitosan is completely dissolved, the solution was stirred for 24 hours at 350 rpm until all the chitosan flake disappear. pH of the solution was adjusted by using 0.1 M NaOH until it reach pH of 5.6 (Guerra et al. 2015) to decrease the acidic condition of chitosan solution. The solution was made up to 100 mL by adding distilled water up to 100mL.

2.2.2. Starch solution.

4% (w/v) of starch solution was gelatinized at 80°C. The solution was stirred for 30 minutes at 350 rpm to overcome the starch from being coagulated. The solution was made up to 100 mL by add up distilled water until reach to 100mL. In order to enhance the application of solution formed by the combining chitosan and CUR essential oil, 2 mL of glycerol was added as a plasticizer.

2.2.3. Chitosan and starch coating solutions.

2% (w/v) of prepared starch solution was added to 100 mL of prepared chitosan solution. The combined solution was stirred at room temperature for 18h to enable the solution well mix. For the combination of EO into chitosan and starch, different concentration of CUR essential oil was used which are 20, 10, 5, and 0 μL. The CUR essential oil was added into the prepared chitosan and stirred at room temperature until complete mixing. The coating solution formulations were added with chitosan concentrations of: 20, 15, 10 and 5 μL; glycerol concentrations of 2.0% and (v/v); Tween 80 concentrations of 0.1% (w/v).

2.3. Characterization of chitosan-starch based coating

The combination of biopolymers give better results compared to only one component (Cao, Shi, and Chen 1998). The importance of surface properties is essential for understanding coating adhesion and to optimize the effectiveness of the coating. In this study, the blending of chitosan and starch coating with different concentration of CUR essential oil has been produced. The coatings were coated on strawberries and then were being evaluated in terms of their surface tension, microstructure and effect to the strawberries.

2.3.1. Surface tension.

In order to examine the adhesion properties of the coating to the strawberry surface, it can be done by measuring the contact angle of a standard liquid on the surface (Moncayo and Buitrago 2015). In this study, analysis of surface tension for all the samples has been done using contact angle meter by the pendant drop method.

2.3.2. Microstructure observation.
The microstructure of coated sample was analysed using a field emission-scanning electron microscopy (FESEM) Zeiss Supra 35VP. The strawberry sample were mounted onto metal stubs and coated with Au using an accelerating voltage of 5kV.

2.3.3. Characterization of strawberries
The laboratory assays were done using fresh strawberries with same the maturity stage, 75% red colour, uniform weight with no sign of mechanical or microbiological injury.

3. Results and discussions

3.1. Surface tension.
Strawberry fruits are very rich in nutrients and have a coarse surface with high moisture conditions which provide a good surface to be coated. The result in Fig.1 shows that surface tension is decreasing as the concentration of the CUR essential oil increased. As reported by Kurek et al. (2013), a monomolecular layer formed by liquid-air interface will reduce the surface tension. Moncayo and Buitrago (2015) in their study have described the meaning of surface tension as the amount of energy needed to increase the surface per area.

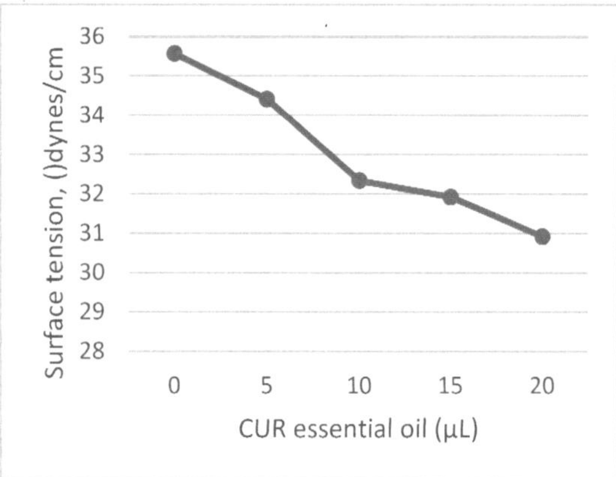

Figure 1. Variation of surface tension of the chitosan-starch based coating with the addition of CUR essential oil

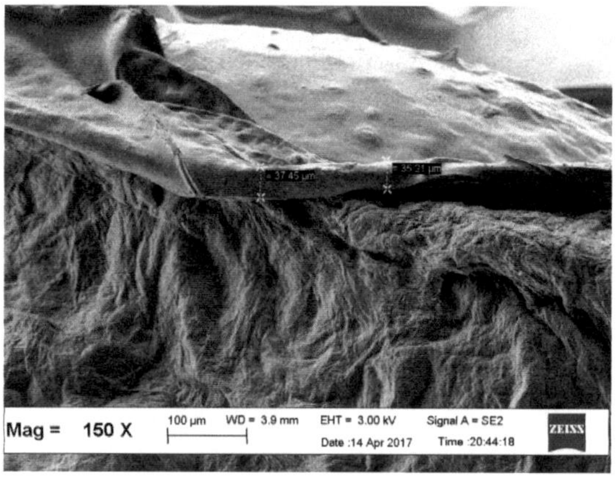

Figure 2. FESEM micrographs of the cross section of the strawberry coated with chitosan-starch based coating containing CUR essential oil

3.2. Microstructure observation.
Field emission scanning electron microscopy (FESEM) was carried out to observe the microstructure and surface characteristics of coating film in order to establish a relationship between coating and strawberry. Fig.2 shows FESEM micrograph of cross section of strawberry coated with chitosan-starch coating plus 15 µL of CUR essential oil. As can be seen the coating had a quite smooth and homogeneous surface with a thickness about 37.45 µm. It suggests that the blending of chitosan and starch with the addition of CUR essential oil are highly miscible and homogeneous.

3.3. Physical appearance

The significance of this study can be seen from results in comparing between coated and uncoated strawberry fruits by qualitative determination through observation. The evolution of infection during the 5th day storage of the coated at different concentration of CUR essential oil and uncoated strawberries is shown in Fig.3 below. On all days of analysis, the strawberries coated with chitosan-starch plus CUR essential oil coatings presented a significantly lower infections than to uncoated. However, the minimum infections was more evident in the strawberries that were coated with 15 µL CUR essential oil in chitosan-starch coating. This is maybe because of too much of CUR essential oil will interfere the antimicrobial agent to be function. Based on study by (Negi 2012), high concentrations of bioactivity compound at lower pH act in a more hydrophobic way. On the other hand, for some other essential oils, they may give a negative impact even at a very low concentration due to their low breakpoint for perception (Lv et al. 2011).

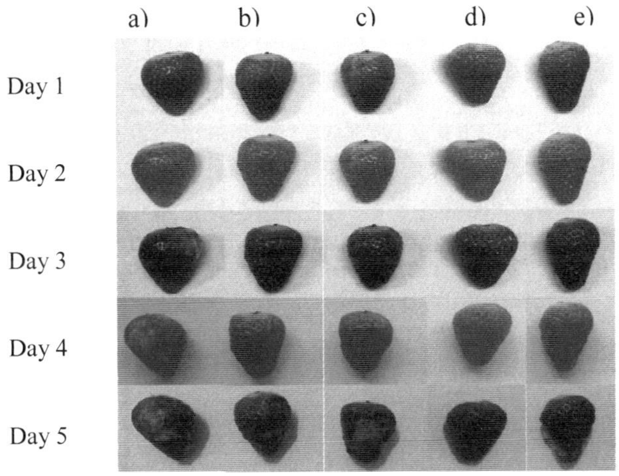

Figure 3. Effect of CUR essential oil in chitosan-starch based coating with different concentrations: a) uncoated, b) 5 µL, c) 10 µL, d) 15 µL and e) 20 µL on strawberry fruits from day 1 until day 5

4. Conclusions

In this study, it was found that application of chitosan-starch based coatings delayed the ripening process and reduced physicochemical changes of strawberry fruit. Chitosan-starch with 15 µL CUR essential oil coatings inhibited decay incidence and maintained until day 5. This study suggested that combine application of chitosan-starch coatings and CUR essential oil can be used for reducing postharvest deterioration and extending the shelf life of strawberry. In addition, the existence of antimicrobial agent such as CUR essential oil had lower down surface tension of the coating. The optimum values of the coatings were obtained experimentally with the formulation of: 1% chitosan, 2 mL of starch, 2% glycerol, 0.1% tween 80 and 15 µL of turmeric oil with surface tension of 31.92 dyne/cm.

Acknowledgments

The authors would like to acknowledge Universiti Teknologi MARA for the financial support to carry out this research through 600-IRMI/MYRA 5/3/LESTARI (0047/2016). Authors also thankful to the staff of Faculty of Chemical Engineering, Universiti Teknologi MARA, Shah Alam for their help in performing laboratory work.

References

[1] Aider, Mohammed. 2010. "Chitosan Application for Active Bio-Based Films Production and

Potential in the Food Industry: Review." *LWT - Food Science and Technology* 43(6): 837–42. http://linkinghub.elsevier.com/retrieve/pii/S002364381000040X.

[2] Azevedo, Alana Novaes et al. 2014. "Response Surface Methodology for Optimisation of Edible Chitosan Coating Formulations Incorporating Essential Oil against Several Foodborne Pathogenic Bacteria." *Food Control* 43: 1–9. http://linkinghub.elsevier.com/retrieve/pii/S0956713514001029.

[3] Cao, Shuguang, Yanqiao Shi, and Guanwen Chen. 1998. "Blend of Chitosan Acetate Salt with poly(N-Vinyl-2-Pyrrolidone): Interaction between Chain-Chain." *Polymer Bulletin* 41(5): 553–59. http://dx.doi.org/10.1007/s002890050401.

[4] Falguera, Víctor et al. 2011. "Edible Films and Coatings: Structures, Active Functions and Trends in Their Use." *Trends in Food Science & Technology* 22(6): 292–303.

[5] Guerra, Ingrid Conceição Dantas et al. 2015. "Coatings Comprising Chitosan and Mentha Piperita L. or Mentha×villosa Huds Essential Oils to Prevent Common Postharvest Mold Infections and Maintain the Quality of Cherry Tomato Fruit." *International journal of food microbiology* 214: 168–78. http://www.sciencedirect.com/science/article/pii/S0168160515300969.

[6] Kurek, Mia et al. 2013. "Carvacrol Affects Interfacial, Structural and Transfer Properties of Chitosan Coatings Applied onto Polyethylene." *Carbohydrate Polymers* 97(1): 217–25.

[7] Lv, Fei, Hao Liang, Qipeng Yuan, and Chunfang Li. 2011. "In Vitro Antimicrobial Effects and Mechanism of Action of Selected Plant Essential Oil Combinations against Four Food-Related Microorganisms." *Food Research International* 44(9): 3057–64. http://dx.doi.org/10.1016/j.foodres.2011.07.030.

[8] Moncayo, Diana, and Gustavo Buitrago. 2015. "The Surface Properties of Biopolymer-Coated Fruit : A Review The Surface Properties of Biopolymer-Coated Fruit : A Review Propiedades de Superficie En Frutas Recubiertas Con Biopolímeros :" (April).

[9] Negi, Pradeep Singh. 2012. "Plant Extracts for the Control of Bacterial Growth: Efficacy, Stability and Safety Issues for Food Application." *International Journal of Food* Microbiology 156(1): 7–17. http://dx.doi.org/10.1016/j.ijfoodmicro.2012.03.006.

[10] Rojas-Graü, María Alejandra, Robert Soliva-Fortuny, and Olga Martín-Belloso. 2009. "Edible Coatings to Incorporate Active Ingredients to Fresh-Cut Fruits: A Review." *Trends in Food* Science & *Technology* 20(10): 438–47. http://linkinghub.elsevier.com/retrieve/pii/S0924224409001769.

[11] Sallato, B. V., R. Torres, J. P. Zoffoli, and B. a. Latorre. 2007. "Effect of Boscalid on Postharvest Decay of Strawberry Caused by Botrytis Cinerea and Rhizopus Stolonifer." *Spanish Journal of* Agricultural *Research* 5(1): 67–78.

[12] Šamec, Dunja et al. 2016. "Assessment of the Differences in the Physical, Chemical and Phytochemical Properties of Four Strawberry Cultivars Using Principal Component Analysis." *Food* Chemistry 194: 828–34.

[13] Santos, N.S.T et al. 2012. "Efficacy of the Application of a Coating Composed of Chitosan and Origanum Vulgare L. Essential Oil to Control Rhizopus Stolonifer and Aspergillus Niger in Grapes (Vitis Labrusca L.)." *Food Microbial* (32): 345–53.

[14] Vásconez, María B. et al. 2009. "Antimicrobial Activity and Physical Properties of Chitosan-Tapioca Starch Based Edible Films and Coatings." *Food Research International* 42(7): 762–69. http://dx.doi.org/10.1016/j.foodres.2009.02.026.

IOP Conf. Series: Journal of Physics: Conf. Series **885** (2017) 012009 doi:10.1088/1742-6596/885/1/012009

Online particle size distribution estimation of a mixture of similar sized particles with acoustic emissions

Ejay Nsugbe, Andrew Starr, Ian Jennions and Cristobal Ruiz Carcel

Cranfield University, College Rd, Cranfield , MK43 0AL

e.nsugbe@cranfield.ac.uk

Abstract. Particle processing plants regard the Particle Size Distribution (PSD) as a key quality factor as it influences the bulk and flow properties of the particles. In this work, Acoustic Emission (AE) is used to estimate the PSD of a mixture that comprise of similar sized particles. The experiments involved the use of regular sized particles (glass beads) and with the aid of a time domain based threshold analysis of the particle impacts the PSD of the mixtures could be estimated.

1. Introduction and related works summary

The Particle Size Distribution (PSD) is an important quality attribute of various kinds of particle processing settings that range from granulators, batch agglomeration mixers, coal milling plants, fluidised beds and pneumatically conveyed systems.[1-5] The need for particle size monitoring methods has seen several technologies developed, some of which have been used in industrial based settings while others are still limited controlled laboratory based systems.[4]

Notable AE based particle sizing began with the work of Leach and Buttle, who observed that a linear relationship exists between particle sizes and their respective AE amplitude.[7][8] The work done by Buttle et al in a vacuum tube assisted in the validation of Leach's theory and provided a mathematical representation of this principle which is referred to as the signal shaping chain and can be seen in figure 1.[8] Buttle quantitatively sized particles using a quantitative sizing technique but due to the deconvolution of each impact peak associated with the signal analysis, this method would be unsuitable to industrial cases where the particle events overlap each other.[2,5,8]

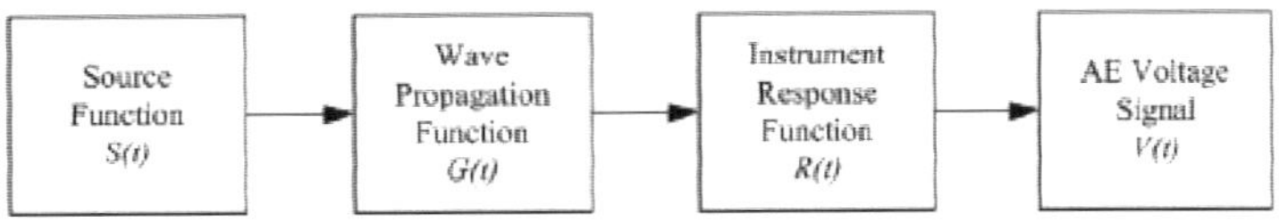

Figure 1. Signal shaping chain **[5][19]**

Data driven techniques were used by Bastari and Chen to estimate particle sizes, due to their approach being structured around neural networks a complete understanding of the decision making process is not fully understood.[3] Hu and Ren applied a threshold and wavelet based methods respectively to estimate particle sizes. Hu et al's method was unable to accurately estimate the sizes of particle under

90 microns while Ren et al's method was seen to not be adaptive enough to work in processes where the final powder mix ratio varies significantly. [4,5]
This paper details the results of an experimental study on a research based on the estimation of the PSD of powder mixtures with AE using a novel signal processing method. [2,9,12] With the goal of this experiment in this report being to observe the performance of the previously designed signal processing method in estimating the PSD of powder mixtures that comprise of similar sized particles.

2. Signal processing approach and experimental setup

The acquired signal will be analysed with analysed with a multiple threshold signal processing method as designed in our previous study. [2,9,12] The principle for the designed multiple threshold method is stemmed from the dynamics of the signal shaping chain, where each particle distribution would have a set of AE amplitudes which can be correlated to the particles themselves.
The same experimental rig as detailed in previous study would be used in this paper, the rig is based around the free flow of powders dispensed at a known rate impelled against the force of gravity on a target plate. [2,9,12]

3. Results and discussion

3.1. Glass beads mixture

Two sets of glass beads were used for the experiments in this section and were mixed in different mix ratios, details of their respective physical properties can be seen in Table 1.

Table 1. Glass beads physical properties

Particle Class	Particle Size Distribution(microns)	Density(g/cm^3)	% Density Difference
Class1	150-212	1.49	2.6%
Class2	212-300	1.53	

Each powder mix ratio was repeated 5 times and a constant mass of powder was dispensed into the experimental funnel each time while the data was analysed using the signal processing method used in previous study.[12] Using the rules set out in the designed signal processing approach, the optimal correlation plot which was chosen as the PSD estimation model can be seen in figure 2. The best correlation plot showed a linear relationship existed between the different particle mixtures and their associated amplitude mean values. The accuracy of this linear model was investigated and a set of particle mixtures were mixed, dispensed and analysed to serve as the model validation mixtures. For the validation of the particle sizing model, each set of AE validation dataset was analysed with the amplitude threshold parameters which produced the best correlation plot, from this this threshold region the mean AE amplitude was extracted and this value is then correlated with its respective particle size using the final chosen correlation plot.

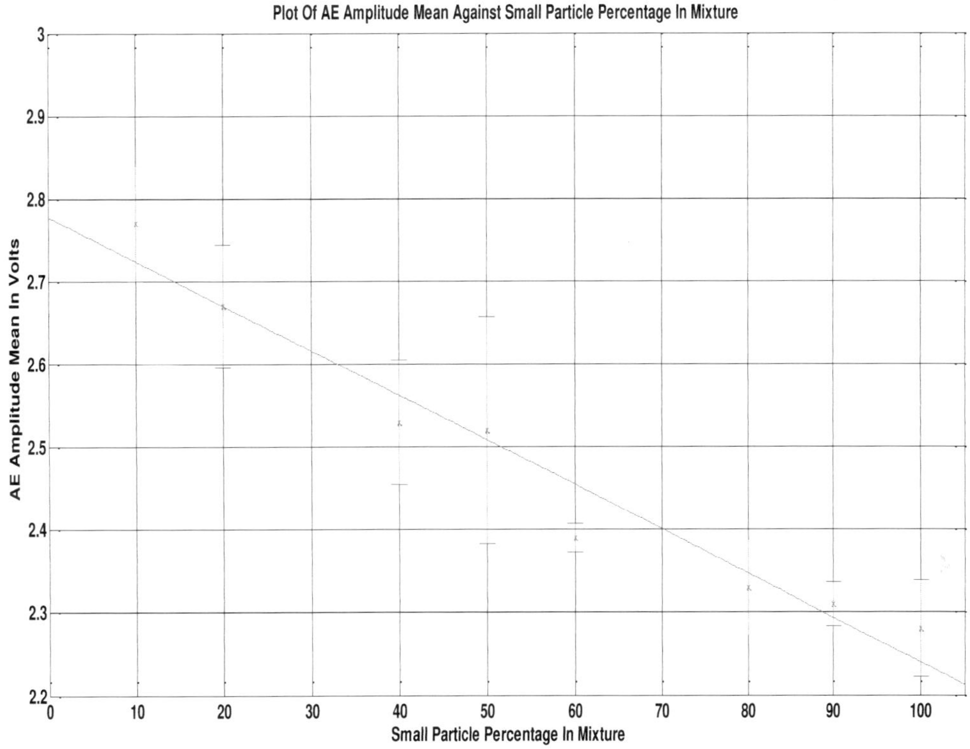

Figure 2. Correlation plot of ae amplitude mean against small particle (Class 1) percentage in mixture

The results of the linear model validation exercise can be seen in figure 3, the model was able to estimate particle mix ratio with an average absolute error of 12%. This would suggest that a good estimation of particle size can be achieved for mixtures which comprise of particles with similar sizes, using the designed multiple threshold signal processing method.

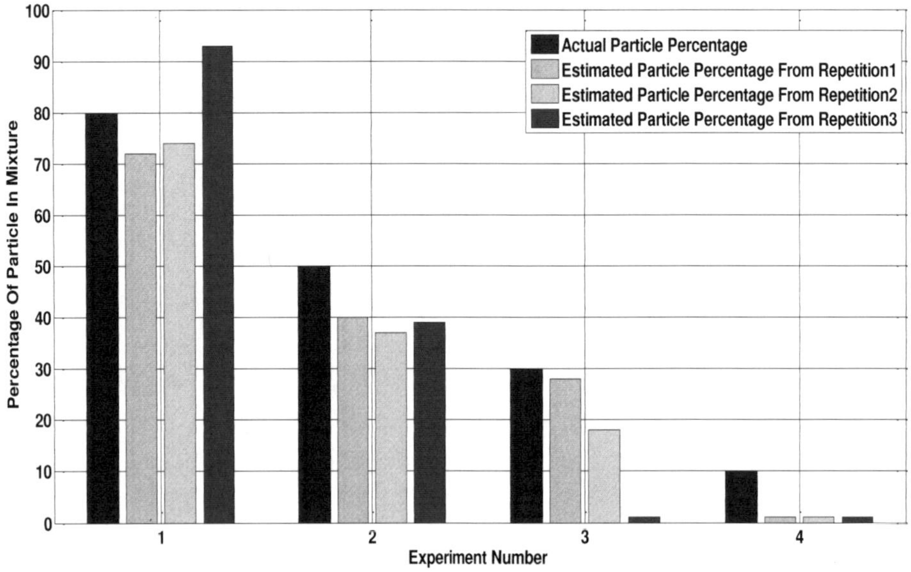

Figure 3. Chart comparing percentage of class 1 glass beads particle against amount estimated by model

4. Conclusion

This work investigated the use of AE to estimate the PSD of mixtures of powders with similar sized particles. The signal processing method used is based on a hybrid PSD estimation method designed in a previous study. [2,9,12] The experiments carried out involved a mixture of two sets of regular sized glass beads whose percentage bulk density difference between them was 2.6%. With the particle bulk density noted as an important factor which influences the output amplitude of a particle during impact. [13]

The result from the experiments conducted showed that the signal processing approach was capable of estimating PSD of the glass beads mixture to within 12% average absolute error. It should be noted that although the chart in figure 2 appears to be estimating only the class 1 particles, the operating assumption is that hence it is a two particle mixture if the amount of one particle group can be estimated with the particle estimation model, the amount of the second particle in the mixture can then be simply calculated. The next set of experiment in this study would involve a mixture comprising of a set of regular and irregular sized particles to observe their behaviour when in a mixture and also in order to quantify the accuracy of the signal processing method.

Acknowledgments

The authors gratefully acknowledge the support of partners Procter & Gamble, IIT, Ajax, Centre for Process Innovation, the University of Birmingham, the University of Leeds, and the University of Durham. This research is part of Chariot and The Advance Manufacturing Supply Chain Initiative (AMSCII) which is a government supply chain fund which is helping to rebuild British manufacturing prowess.

References

[1] Papp, M.K., Pujara, C.P., Pinal, R.: Monitoring of high-shear granulation using acoustic emission: predicting granule properties, J Pharm. Innov., 2008, 3, pp. 113–122

[2] E.Nsugbe A.Starr C.Ruiz-Carcel Monitoring The Particle Size Distribution Of A Powder Mixing Process With Acoustic Emissions: A Review. Engineering Technology Reference,

pp. 1–12 ,doi: 10.1049/etr.2016.0139
[3] Bastari, A., Cristalli, C., Morlacchi, R.,: Acoustic emissions for particle sizing of powders through signal processing techniques, Mech. Syst. Signal Process., 2011, 25, (3), pp. 901–916
[4] Ren, C.J., Wang, J.D., Song, D., et al.: Determination of particle size distribution by multi-scale analysis of acoustic emission signals in gas-solid fluidized bed, J. Zhejiang Univ., Sci A, 2011, 12, (4), pp. 260–267
[5] Hu, Y., Huang, X., Qian, X., et al.: Online particle size measurement through acoustic emission detection and signal analysis. Proc. IEEE Int. Instrumentation Measurement Technology Conf., May 2014, pp. 949–953
[6] Endecotts layer sieve website accessed on 12/12/17 http://www.endecotts.com/products/sieve-shakers/
[7] Leach, M.F., Rubin, G.A., Williams, J.C.: 'Particle size determination from acoustic emissions', Powder Technol., 1977,16, pp. 153–158
[8] D.J. Buttle, S.R. Martin, C.B. Scruby, Particle sizing by quantitative acoustic emission, Research in Nondestructive Evaluation 3 (1991) 1–26.
[9] E Nsugbe, A Starr, I Jennions and C Ruiz-Carcel Online Particle Size Distribution Estimation Of A Particle Mixture In Free Fall With Acoustic Emission(being reviewed by the Journal Of Meas. Sci. and Instrumentation)
[10] Ivantsiv, V., Spelt, J., Papini, M.: 'Mass flow rate measurement in abrasive jets using acoustic emission', Meas. Sci. Technol., 2009, 20, pp. 095402
[11] Pecorari, C.: 'Characterizing particle flow by acoustic emission', J. Nondestruct. Eval., 2013, 32, (1), pp. 104–111
[12] E Nsugbe, A Starr, P Foote, C Ruiz-Carcel and I Jennions Size Differentiation Of A Continuous Stream Of Particles Using Acoustic Emissions IOP Conf. Series: Materials Science and Engineering 161 (2016) 012090 doi:10.1088/1757-899X/161/1/012090
[13] L. Gao, Y. Yan, R. M. Carter, D. Sun, P. Lee and C. Xu, "On-line particle sizing of pneumatically conveyed biomass particles using piezoelectric sensors," Fuel, vol. 113, no. 11, pp. 810-816, 2013.

IOP Conf. Series: Journal of Physics: Conf. Series **885** (2017) 012010 doi:10.1088/1742-6596/885/1/012010

Multi objective Taguchi optimization approach for resistance spot welding of cold rolled TWIP steel sheets

Mumin Tutar[1], Hakan Aydin[1], Ali Bayram[1]

[1]Uludağ University, Faculty of Engineering, Department of Mechanical Engineering, 16059, Görükle-Bursa / TURKEY

mumintutar@uludag.edu.tr

Abstract. Formability and energy absorption capability of a steel sheet are highly desirable properties in manufacturing components for automotive applications. TWinning Induced Plastisity (TWIP) steels are, new generation high Mn alloyed steels, attractive for the automotive industry due to its outstanding elongation (%40-45) and tensile strength (~1000MPa). So, TWIP steels provide excellent formability and energy absorption capability. Another required property from the steel sheets is suitability for manufacturing methods such as welding. The use of the steel sheets in the automotive applications inevitably involves welding. Considering that there are 3000-5000 welded spots on a vehicle, it can be interpreted that one of the most important manufacturing method is Resistance Spot Welding (RSW) for the automotive industry. In this study; firstly, TWIP steel sheet were cold rolled to 15% reduction in thickness. Then, the cold rolled TWIP steel sheets were welded with RSW method. The welding parameters (welding current, welding time and electrode force) were optimized for maximizing the peak tensile shear load and minimizing the indentation of the joints using a Taguchi L9 orthogonal array. The effect of welding parameters was also evaluated by examining the signal-to-noise ratio and analysis of variance (ANOVA) results.

1. Introduction

The combination of strength and ductility of a steel is the most important property for the development of lightweight vehicles with lower fuel consumption and higher safety. As an advanced high strength steels (AHSS), Twinning Induced Plasticity (TWIP) steels with high manganese austenitic microstructure and high strain hardening capacity offer to fulfill this request. Early work on TWIP steels, which is discovered by Hadfield in 1988, was carried out late 1900's [1–4]. There are numerous studies about microstructural and mechanical properties, effect of alloying elements, strain hardening mechanism, heat treatments etc. of TWIP steels [3,5–8].

On the other hand, suitability for manufacturing process is another important property which materials should provide. For the automotive industry, welding, especially resistance spot welding (RSW), is a mandatory process for almost all body parts. Consequently, welding of TWIP steels became a significant issue for usage in automotive industry. Some researchers investigated laser welding (LW) [9–12], dissimilar welding [4,11,13], friction stir spot welding[14] and RSW [15–19] of TWIP steels. However, the effect of the welding parameters on the mechanical and microstructural properties and the optimization of the welding parameters for TWIP steels is not well studied.

IOP Conf. Series: Journal of Physics: Conf. Series **885** (2017) 012010 doi:10.1088/1742-6596/885/1/012010

Almost all studies were focused on only load bearing capacity of the welded joints. But, not only mechanical properties but also indentation has a vital importance for automotive industry. In this study, a multi objective optimization approach was conducted using tensile shear load and indentation values for RSW of 1.3 mm 1Gpa grade and %15 cold rolled TWIP steel sheets. The mechanical properties of the weld were examined by means of TST and micro hardness tests. Effects of the welding parameters on the RSWed joints were discussed.

2. Material and method

In this study, as-received (1.3 mm) TWIP steel sheets were cold rolled %15 in thickness (around 1.1 mm) using cold rolling. All experiments were conducted on %15 cold rolled TWIP steel sheets. The chemical composition of the sheets used in this study is shown in Table 1.

Table 1. The chemical compositions (wt.%) and mechanical properties of %15 cold rolled TWIP steel sheets used in this investigation.

Steel	Fe	C	Mn	Si	Al	Cr	Ti	Ultimate Tensile Strength (MPa)	Total Elongation (%)
TWIP	*Balance*	0.28	15.6	1.06	1.89	0.564	0.1	1220	20

TWIP steel sheets were resistance spot welded in the overlapped configuration using MFDC resistance spot-welding machine connected to ABB robot arm and copper alloy electrodes having the face diameter of 6 mm. Prior to welding experiments, in order to determine the range of focused process parameters for RSW (welding current, welding time and electrode force), preliminary tests were carried out. The accepted criteria for the preliminary tests were the smallest button in peeling test and a maximum %35 indentation value.

Taguchi approach based multi objective optimization was performed using L9 orthogonal array and the parameter values shown in Table 2. First, S/N ratios were calculated separately for maximizing the tensile shear load (TSL) and minimizing the indentation values. In this step, calculated S/N values were normalized using the following equation:

$$Normalized\ S/N = \frac{S/N_i - S/N_{min}}{S/N_{max} - S/N_{min}}$$

where S/Ni is the calculated S/N ratio for the ith performance characteristics.

Then, the values used in Taguchi optimization were calculated by averaging normalized S/N values of TSL and indentation. To determine the relative effect of the parameters ANOVA procedure was also performed using the normalized S/N ratio values.

3. Results and discussion

The tensile shear load and indentation values of welded samples were presented with calculated S/N ratios in Table 2. The calculated heat input index values were also shown in this Table. Additionally, to determine the influence of each level of the response tables were calculated and shown in Table 3. Delta values of the welding parameters shows the magnitude of the effect on response values and rank values shows the order of importance of the welding parameters. Table 3 indicates that the most influential welding parameter was welding current. Main effects plot for the normalized S/N ratios plotted using the values in Table 3 (Fig. 1). Predicted optimum levels of welding parameters were also indicated using asterisk.

IOP Conf. Series: Journal of Physics: Conf. Series **885** (2017) 012010 doi:10.1088/1742-6596/885/1/012010

Table 2. Welding parameters and their TSL, indentation, heat input index values and S/N ratios.

Sample No.	Electrode Force (N)	Weld Current (kA)	Welding Time (ms)	Heat Input Index*	TSL [kN]	Indentation [%]	Average of Normalized S/N Ratio of TSL and Indentation
1	1500	3	100	600	4.18	1.39	0.522
2	1500	6	250	6000	9.71	16.92	0.530
3	1500	9	400	21600	10.54	35.21	0.475
4	2250	3	250	1000	4.77	1.31	0.558
5	2250	6	400	6400	10.23	12.09	0.623
6	2250	9	100	3600	10.61	28.90	0.484
7	3000	3	400	1200	3.96	1.67	0.463
8	3000	6	100	1200	8.56	14.44	0.511
9	3000	9	250	6750	11.61	29.55	0.527

*Heat Input Index = I^2t/F (I: Weld current (A); t: Welding time(s); F: Electrode force (N))

Table 3. Response table for Normalized S/N Ratios.

	Means		
Level	**Electrode Force**	**Weld Current**	**Welding Time**
1	0.509	0.515	0.506
2	0.555*	0.555*	0.538*
3	0.500	0.495	0.520
Delta	0.055	0.059	0.033
Rank	2	1	3

* Optimum level

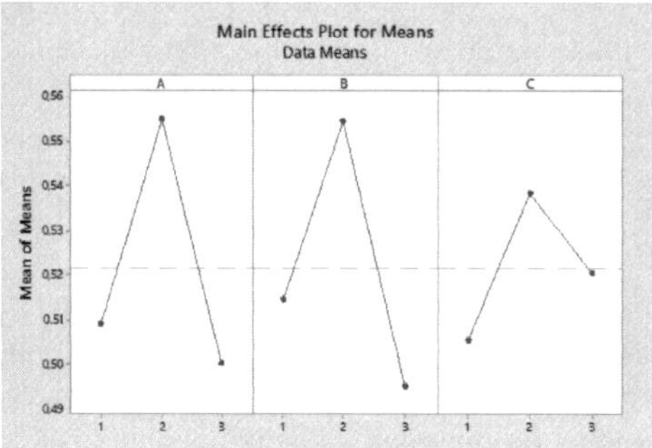

Figure 1. Main effects plot for the Normalized S/N Ratios.

According to Taguchi calculations, optimum levels of the welding parameters found to be 2250 N for welding force, 6 kA for welding current and 250 ms for welding time. Predicted optimum parameter set was not in the used orthogonal array, and thus, confirmation experimental test was required. To confirm the optimum levels of welding parameters, joining process was carried out again using predicted optimum parameter set and tensile shear load – deformation curves of selected samples having minimum, maximum and mean heat input index and optimum sample were presented in Fig 2. Obtained results of confirmation test were given in Table 4. The obtained maximum TSL was 11.61 kN (sample 9) with %29.55 indentation. Despite, %9.6 decrease in TSL, %59 decrease in indentation was achieved by optimization of welding parameters.

To determine the relative effect of the welding parameters and compare with Taguchi results, the standard ANOVA procedure was also performed using the normalized S/N ratios(Table 5). The most important parameter defined as weld current with a contribution of 29.22 percent. The order of importance was also in a good relationship with Taguchi result.

Table 4. Experimental results of confirmation test.

Electrode Force (N)	Weld Current (kA)	Welding Time (ms)	TSL (kN)	Indentation (%)
2250	6	250	10.49	12.00

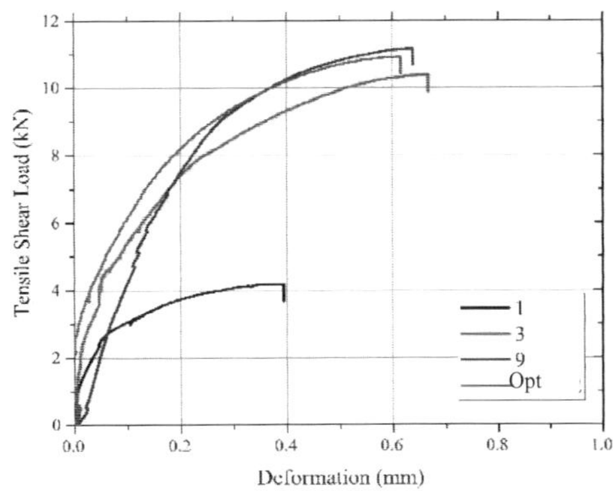

Figure 2. Tensile shear load – Deformation curves of selected samples.

Table 5. ANOVA table for normalized S/N ratios.

Source	DOF	Seq SS	Adj SS	Adj MS	F	Contribution (%)
Electrode Force	2	0.0052	0.0052	0.0026	0.79	27.51
Weld Current	2	0.0055	0.0055	0.0028	0.84	29.22
Welding Time	2	0.0016	0.0016	0.0008	0.25	8.57
Residual Error	2	0.0066	0.0066	0.0033		34.69
Total	8	0.0189				100.00

4. Conclusions

In this study, RSW welding parameters of cold rolled TWIP sheet steel joints have been multi-objectively optimized using Taguchi approach. Also, the effects of welding parameters were discussed by means of TST and indentation values of the joints. The main conclusions can be drawn as follows:

- Optimum levels of the welding parameters found to be 2250 N welding force, 6 kA welding current and 250 ms welding time.
- The order of importance of the welding parameters was found to be: weld current > electrode force > welding time.
- The sample welded using optimum parameter set showed %9.6 lower TSL and %59 lower indentation than the sample which has maximum TSL.

Acknowledgments

The authors are grateful to the Scientific and Technological Research Council of Turkey (TUBITAK) for its financial support to this research (Project number: MAG 213M597). The authors are also grateful to Ermetal Inc. for providing facilities for the resistance spot-welding processes.

References

[1] N. Zan, H. Ding, X. Guo, Z. Tang, W. Bleck, Effects of grain size on hydrogen embrittlement in a Fe-22Mn-0 . 6C TWIP steel, Int. J. Hydrogen Energy. 40 (2015) 1–10. doi:10.1016/j.ijhydene.2015.06.112.

[2] L. Halbauer, R. Zenker, A. Weidner, A. Buchwalder, H. Biermann, Electron Beam Welding of

Cold Rolled High-Alloy TRIP/TWIP Steel Sheets, Steel Res. Int. (2015) n/a–n/a. doi:10.1002/srin.201500086.

[3] Y. Zhao, J. Wang, S. Zhou, X. Wang, Effects of rare earth addition on microstructure and mechanical properties of a Fe–15Mn–1.5Al–0.6C TWIP steel, Mater. Sci. Eng. A. 608 (2014) 106–113. doi:10.1016/j.msea.2014.04.084.

[4] P. Russo Spena, M. De Maddis, F. Lombardi, M. Rossini, Dissimilar Resistance Spot Welding of Q&P and TWIP Steel Sheets, Mater. Manuf. Process. 31 (2016) 291–299. doi:10.1080/10426914.2015.1048476.

[5] T. Dieudonné, L. Marchetti, M. Wery, J. Chêne, C. Allely, P. Cugy, et al., Role of copper and aluminum additions on the hydrogen embrittlement susceptibility of austenitic Fe-Mn-C TWIP steels, Corros. Sci. 82 (2014) 218–226. doi:10.1016/j.corsci.2014.01.022.

[6] K. Renard, P.J. Jacques, On the relationship between work hardening and twinning rate in TWIP steels, Mater. Sci. Eng. A. 542 (2012) 8–14. doi:10.1016/j.msea.2012.01.123.

[7] J. Hajkazemi, a. Zarei-Hanzaki, M. Sabet, S. Khoddam, Double-hit compression behavior of TWIP steels, Mater. Sci. Eng. A. 530 (2011) 233–238. doi:10.1016/j.msea.2011.09.080.

[8] T. Declara, I. Cortes, N.A.S. Pens, E.S.D.E. Aposenta, D.E. Sobreviv, P.P. Cga, et al., Effect of heat treatment and hot working on the microstructural characteristics of TWIP steels, (2014) 1–81.

[9] L. Mujica, S. Weber, H. Pinto, C. Thomy, F. Vollertsen, Microstructure and mechanical properties of laser-welded joints of TWIP and TRIP steels, Mater. Sci. Eng. A. 527 (2010) 2071–2078. doi:10.1016/j.msea.2009.11.050.

[10] T. Wang, M. Zhang, W. Xiong, R. Liu, W. Shi, L. Li, Microstructure and tensile properties of the laser welded TWIP steel and the deformation behavior of the fusion zone, Mater. Des. 83 (2015) 103–111. doi:10.1016/j.matdes.2015.06.002.

[11] V. Behm, M. Höfemann, A. Hatscher, A. Springer, S. Kaierle, D. Hein, et al., Investigations on Laser Beam Welding Dissimilar Material Combinations of Austenitic High Manganese (FeMn) and Ferrite Steels, Phys. Procedia. 56 (2014) 610–619. doi:10.1016/j.phpro.2014.08.049.

[12] H. Li, H. Jiang, L. Yang, Y. Wang, Mechanical properties and microstructure of laser welded TWIP steels, Cailiao Kexue Yu Gongyi/Material Sci. Technol. 22 (2014) 6–9. http://www.scopus.com/inward/record.url?eid=2-s2.0-84925271057&partnerID=tZOtx3y1.

[13] P.R. Spena, P. Matteis, A. Sanchez, G. Scavino, Strength and fracture of TWIP steel dissimilar weld joints, Convegno IGF XXII Roma. (2013) 109–117. http://www.gruppofrattura.it/ocs/index.php/cigf/IGF22/paper/view/10900.

[14] M.M.Z. Ahmed, E. Ahmed, A.S. Hamada, S.A. Khodir, M.M. El-Sayed Seleman, B.P. Wynne, Microstructure and mechanical properties evolution of friction stir spot welded high-Mn twinning-induced plasticity steel, Mater. Des. (2015). doi:10.1016/j.matdes.2015.12.001.

[15] M.H.H. Razmpoosh, M. Shamanian, M. Esmailzadeh, The microstructural evolution and mechanical properties of resistance spot welded Fe – 31Mn – 3Al – 3Si TWIP steel, Mater. Des. 67 (2015) 571–576. doi:10.1016/j.matdes.2014.10.090.

[16] D.C. Saha, S. Han, K.G. Chin, I. Choi, Y. Do Park, Weldability evaluation and microstructure analysis of resistance-spot- welded high-Mn steel in automotive application, Steel Res. Int. 83 (2012) 352–357. doi:10.1002/srin.201100324.

[17] D.C. Saha, Y. Cho, Y. Park, Metallographic and fracture characteristics of resistance spot welded TWIP steels, 18 (2013) 711–720. doi:10.1179/1362171813Y.0000000151.

[18] P. Russo Spena, M. De Maddis, F. Lombardi, M. Rossini, Investigation on Resistance Spot Welding of TWIP Steel Sheets, Steel Res. Int. 86 (2015) 1480–1489. doi:10.1002/srin.201400336.

[19] J. Yu, J. Shim, S. Rhee, Characteristics of Resistance Spot Welding for 1 GPa Grade Twin Induced Plasticity Steel, Mater. Trans. 53 (2012) 2011–2018. doi:10.2320/matertrans.M2012167.

Investigation on heat transfer characteristic and optimization of the cooling air inlet for the twin-web turbine disk

Mengchuang ZHANG*, Wenxuan GOU, Qin YAO, Lei Li, Zhufeng YUE

Department of Engineering Mechanics, Northwestern Polytechnical University (Chang'an Campus), Xi'an, P.R. China.

*zmc@mail.nwpu.edu.cn,

Abstract. With a higher operation temperature, the conventional aero-turbine single web disk (SWD) has reached its limits. The twin-web disk (TWD) has been designed as a breakthrough, which has an expected performance in weight loss, strength and heat transfer efficiency. However, the lack of investigation on the position of the cooling air inlet is slowing down further application of TWD. Therefore, for a further study, inlet position optimization with maximum average Nusselt number is conducted for TWD flow structure study. The average Nusselt number result shows that the TWD has a better performance in heat transfer. All the works, including modeling and analyzing, can be referred for engineering design. And the conclusions obtained in this paper could be valuable for the future improvement of the TWD.

1. Introduction

Currently, higher turbine inlet temperature (TIT) has reached 1600K and expected to be pushed toward 2000K in the future, which is beyond the limits of the most parts of aero-engine turbines, especially the first stage turbine (i.e., high pressure turbine, or HPT) disk [1]. The conventional aero-turbine single web disk (SWD) has reached its limit of the AN2 (turbine annulus area multiplied by speed squared, unit $m^2(r/min)^2$) and thermal loads. As a breakthrough, the twin-web disk (TWD) has been designed with an expected performance in weight loss, strength and heat transfer efficiency. Therefore, it has been proven to be the future trend of the high pressure turbine disk (HPT) by the U.S. IHPTET program [2, 3].

Many researchers have studied on the SWD cavity and its discoidal rotor-stator systems using experimental or numerical methods [4-6]. Harmand et al. [7] reviewed many convinced numerical technologies during the past decades. In the reference [8-10], the decoupled and conjugate methods are widely used. Harmand [7] indicated that the conjugate heat transfer method is an efficient and accurate coupling analysis numerical method and has been widely used in the numerical study on the heat transfer characteristics of gas turbine disk [11-13] and other parts of the gas turbine engine [14-16] with a desired accuracy.

In recent years, optimization based on thermo-fluid analysis is becoming increasingly popular in engineering design [17, 18] and turbine disk[19]. In some cases, evolutionary algorithms are used to ensure reaching the global optimum. Besides, thermo-fluid analysis often includes the solution of high nonlinear equations, the cost of time is huge and unacceptable. As a lower fidelity computational model, two-dimensional model is used widely in aerodynamic optimization [20, 21].

Therefore, the two-dimensional model and structural mesh, which can reduce the cost of the optimization searching by a large margin, are used to find the optimum cooling air inlet position. For the purpose of comparing with the SWD in [22], the models in this study have the same size and boundary conditions. All the works, including modeling and analyzing, can be referred for engineering design.

2. Numerical method

2.1. Geometry model and grid

The 2D dimensional model (fluid region) is shown in Fig. 1 accordingly to Ding's experiment [22]. It is unable to conduct the calculation of 2D model by ANSYS CFX directly. A thin model with only single layer grid in z direction is often used as a solution. Structural grid is also shown in Fig. 1. To ensure that solutions yield sufficient accuracy within CFX, a mesh dependency study has been performed. The average total pressure value and the average temperature are set as the baseline for the mesh independence study based on reference [23]. It is found that the mesh constitutes a satisfying compromise between the duration and the accuracy of the calculation for optimization purpose.

The heat transfer on the rim surface consists of two parts: the heat conduction from the heated blade and the heat convection from the high temperature gas. A constant and uniform heat flux is applied here. The detailed boundary conditions are listed in table 1. The layers of near wall mesh are defined as 7 to satisfy the requirement.

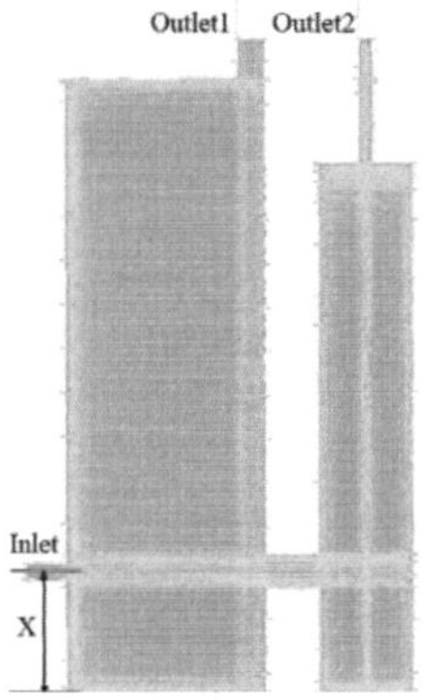

Figure 1. 2D dimensional model for optimization

Table 1. Boundary condition of the calculation model

	Parameters	Symbols	Given values
Working	Uniform stress on disk rim	PRES_RIM	340 MPa
condition	Working rotating speed	RPM	14000 rev / min
Thermo-fluid	Temperature of cooling air	T_IN	600 K
boundary	Temperature at rim	T_OUT	850 K
condition	Mass flow rate of inlet	MFLOW_INLET	500 kg/h
	Average static pressure of outlet1	P_OUTLET1	6 atm
	Average static pressure of outlet2	P_OUTLET2	8atm

2.2. Turbulence models

The SST turbulence model is considered a good quality in interface data transition and therefore offers a good compromise between accuracy and calculation cost [7]. In order to obtain a reasonable computational result, Y+ is set in a range from 10 to 30 based on the near-wall grid requirement by SST model. Rotational Reynolds number is defined as:

$$\mathrm{Re}_{\omega} = \rho\omega R^2/\mu \quad (1).$$

Where ρ is the density of the air, ω is the rotational speed of the disk, R is the radius, μ is the dynamic viscosity of the air.

2.3. Experiments verification

Shuiting et al. [24] have conducted an experiment to investigate the flow and heat transfer on rotor-stator system with pre-swirl angle. Fig.2 shows the configuration of a SWD in the experiment. The data of the SWD sample can be easily obtained from the reference. 1/6 of the rotational computational model in 6 cases (mass flow rate from 200kg/h to 450kg/h) are analyzed for heat transfer by the numerical method above. The numerical result for the temperature distribution of computational SWD ($\dot{m} = 450kg/h$) is shown in Fig.3.

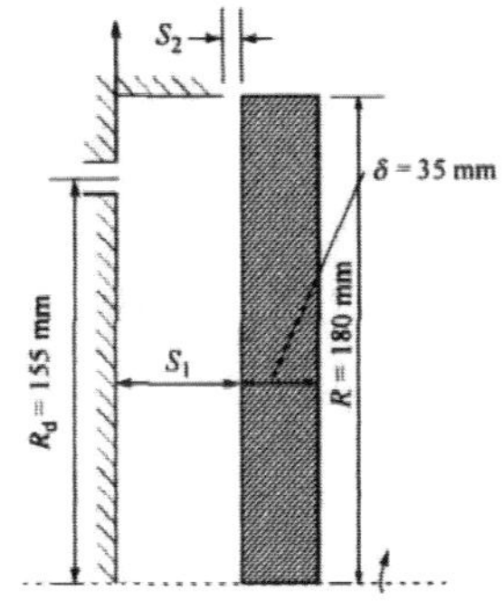

Figure 2. Schematic configuration of a TWD in Shuiting's experiment

Fig.4 shows the numerical and experimental results in all the cases. The most difference between the numerical and experimental results is the region of disk hub in the case 1 ($\dot{m} = 200kg/h$). Case 6 ($\dot{m} = 200kg/h$) is the most accurate model with the only 1K of the maximum deviation. Therefore, larger mass flow rate ($\dot{m} = 500kg/h$) is used in this paper in order to get expected accuracy.

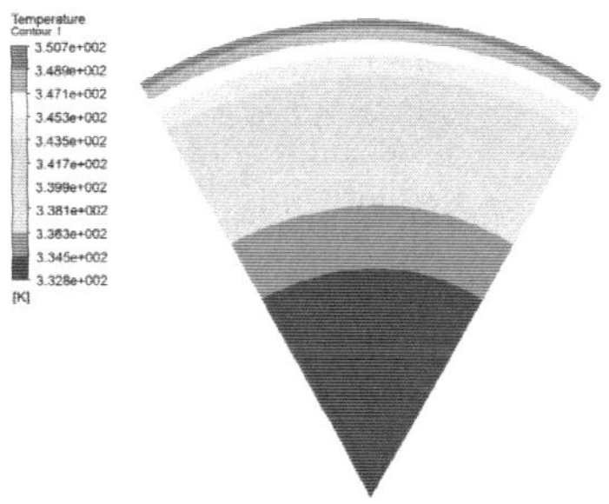

Figure 3. The temperature distribution of computational SWD ($\dot{m}$ = 450kg/h)

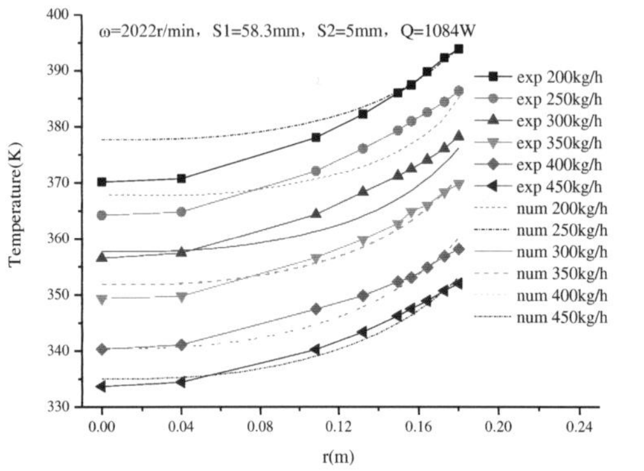

Figure 4. The temperature of the SWD comparison between the computation and experiment

3. Optimization models

3.1. Design variable and object function

The dimensionless inlet position x/R is used set as the design variable in a range of 0~0.9. The object is the maximum average Nusselt number Nu_{ave}:

$$Nu_{ave} = \alpha_{av} R/\lambda \tag{1}$$

Where α_{av} is the average heat transfer coefficient, R is the disk radius, and λ is the thermal conductivity of the static fluid. As the only unknown variable, the average heat transfer coefficient α_{av} can be obtained by the ANSYS CFX. Then the optimization problem can be defined as Maximize Nu_{ave}.

3.2. Optimization technique

Evolution optimization algorithm is a stochastic global search method. The algorithm is operated on a population of potential solutions applying the principle of survival of the fittest to generate better approximations. For each generation, mutation is introduced to reduce the possibility of local optimal search. The evolution strategy here is based on the works of Rechenberg and Schwefel which mutates designs by adding a normally distributed random value to each design variable. The mutation strength (standard deviation of the normal distribution) is self-adaptive and changes during the optimization process [25].

3.3. Optimization process

ISIGHT optimization software with ANSYS CFX are adopted here to find optimum designs. The automatic process control is implemented in the windows batch file and windows executable files generated by FORTRAN. The geometry of the model is changed with the changes of geometrical parameters input by the ISIGHT optimizer.

4. Results

Included mesh generation, CFD analysis, using the 32 core CPU, every optimized step last for about 1.5 min in total. The mean iterations of the computation for thermal-fluid using ANSYS CFX is about 60 and more than 60% of the total time cost. After 51 optimized iterations, the final global optimal results are obtained by evolution optimization algorithm. Fig. 5 shows the optimization history of object and design variables, which can be easily seen the characteristic of the evolution optimization algorithm. The optimum (x/R = 0.432, Nu_{ave} = 2853.1) is also shown in Fig. 12 in green color.

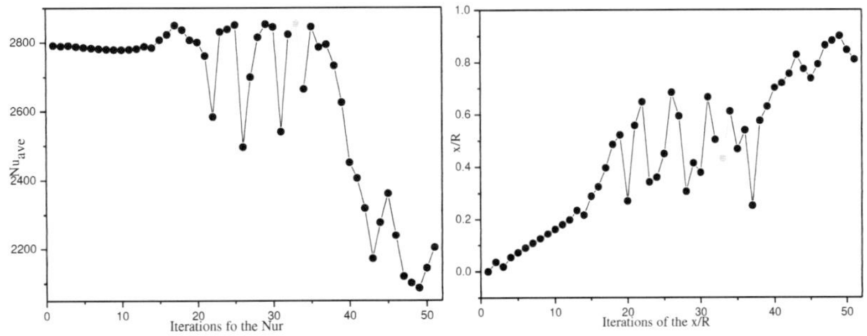

Figure 5. The optimization history of object (a) Nu_{ave} ; (b) dimensionless inlet position x/R

In order to study the sensibility of different inlet positions, the grids and the velocity streamline of 5 models are shown in Fig.6 and 7 respectively. It is obvious that higher inlet position form more vortex in the cavity of TWD, but the velocity loss is also higher. The two aspects brought by the higher inlet position has opposite function on Nusselt number, while the optimum x/R = 0.432 satisfies both aspects.

The Nu_{ave} of TWD with the sequential dimensionless inlet position x/R is plotted in Fig.8, along with the Nu_{ave} data of SWD from the reference [22]. It shows that the optimum inlet positions of SWD and TWD are different. TWD has a higher heat transfer performance in the region of disk hub and a global higher Nu_{ave} in disk surface.

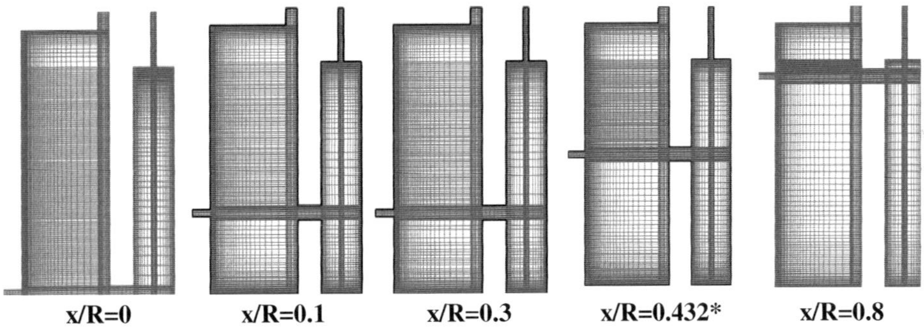

Figure 6. The grids of 5 models with different inlet position (* - the optimum)

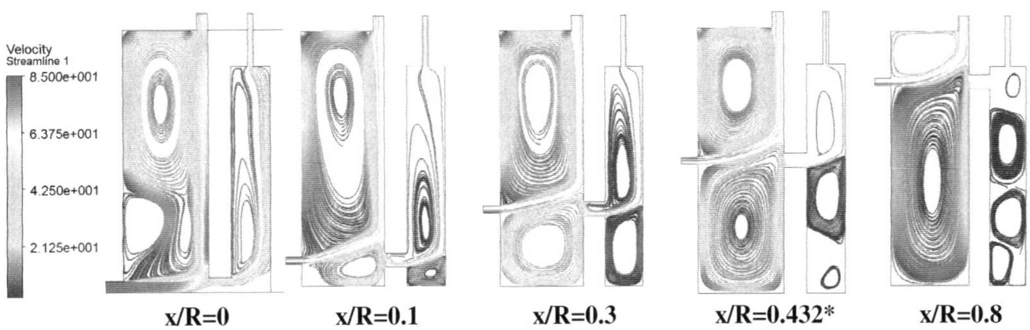

Figure 7. Velocity streamline study of 5 models with different inlet position (* - the optimum)

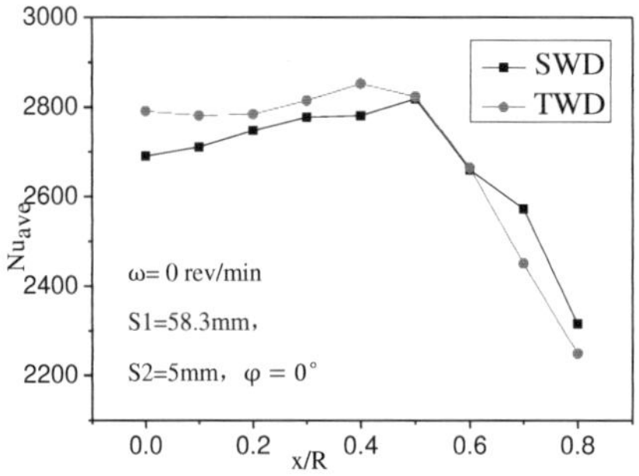

Figure 8. Nu_{ave} comparison between SWD and TWD

5. Conclusions

In this paper, a global optimization is conducted with the object of the maximum average Nusselt number. The results show that x = 0.432 is the best position. The Nu_{ave} of TWD and SWD with the sequential dimensionless inlet position x/R also indicate that TWD has a better performance in heat transfer. The works in this paper could be valuable for the future improvement of the TWD.

Acknowledgment

This work is supported by China Postdoctoral Science Foundation(Grant No. 2014M562281), Nationa l Natural Science Foundation of China (Grant No. 51205315) and Aerospace Technology Support Fou ndation(2014-HT-XGD).

Reference

[1] LIAO G, WANG X, LI J, et al. Numerical investigation on the flow and heat transfer in a rotor-stator disc cavity [J]. Applied Thermal Engineering, 2015, 87(0): 10-23.

[2] CAIRO R R, SARGENT K A. Twin web disk: A step beyond convention [J]. Journal of Engineering for Gas Turbines and Power-Transactions of the Asme, 2002, 124(2): 298-302.

[3] SHEN X L, DONG S J. Structure Optimization and Welding Residual Stress Analysis of Twin-Web Turbine Disc [J]. Advanced Materials Research, 2013, 622-623(

[4] KARABAY H, WILSON M, OWEN J M. Predictions of effect of swirl on flow and heat transfer in a rotating cavity [J]. International Journal of Heat & Fluid Flow, 2001, 22(143-55.

[5] TRAN N, CHANG Y-J, TENG J-T, et al. Numerical and Experimental Investigations on Heat Transfer of Aluminum Microchannel Heat Sinks with Different Channel Depths [J]. International Journal of Mechanical Engineering and Robotics Research, 2015, 4(3): 204.

[6] POUNTNEY O J, SANGAN C M, LOCK G D, et al. Effect of Ingestion on Temperature of Turbine Disks [J]. Journal of Turbomachinery, 2013, 135(5): 1790-1.

[7] HARMAND S, PELL J, PONCET S, et al. Review of fluid flow and convective heat transfer within rotating disk cavities with impinging jet [J]. International Journal of Thermal Sciences, 2013, 67(5): 1–30.

[8] SMITH P E J, MUGGLESTONE J, THAM K M, et al. Conjugate Heat Transfer CFD Analysis in Turbine Disc Cavities; proceedings of the ASME Turbo Expo 2012: Turbine Technical Conference and Exposition, F, 2012 [C].

[9] OKITA Y, YAMAWAKI S. Conjugate Heat Transfer Analysis of Turbine Rotor-Stator System;

proceedings of the ASME Turbo Expo 2002: Power for Land, Sea, and Air, F, 2002 [C].

[10] ANDREI L, ANDREINI A, FACCHINI B, et al. A Decoupled CHT Procedure: Application and Validation on a Gas Turbine Vane with Different Cooling Configurations ☆ [J]. Energy Procedia, 2014, 1087–96.

[11] ANDREINI A, SOGHE R D, FACCHINI B. Turbine Stator Well CFD Studies: Effects of Coolant Supply Geometry on Cavity Sealing Performance [J]. Journal of Turbomachinery, 2009, 133(2): 841-53.

[12] GANINE V, JAVIYA U, HILLS N, et al. Coupled Fluid-Structure Transient Thermal Analysis of a Gas Turbine Internal Air System With Multiple Cavities [J]. Journal of Engineering for Gas Turbines & Power, 2012, 134(10): 2167-77.

[13] SUN Z, CHEW J W, HILLS N J, et al. Coupled Aerothermomechanical Simulation for a Turbine Disk Through a Full Transient Cycle [J]. Journal of Turbomachinery, 2011, 134(1): 112-21.

[14] HEIDMANN J D, KASSAB A J, DIVO E A, et al. Conjugate Heat Transfer Effects on a Realistic Film-Cooled Turbine Vane; proceedings of the ASME Turbo Expo 2003, collocated with the 2003 International Joint Power Generation Conference, F, 2003 [C].

[15] KIM K M, PARK J S, DONG H L, et al. Analysis of conjugated heat transfer, stress and failure in a gas turbine blade with circular cooling passages [J]. Engineering Failure Analysis, 2011, 18(4): 1212–22.

[16] DING S, LI G, LUO B. Active Control Thermal-Loading Method to Ameliorate Stress in Aeroengine Turbine Disk [J]. Journal of Thermophysics & Heat Transfer, 2013, 27(2): 274-85.

[17] ALEXANDROV N M, LEWIS R M, GUMBERT C R, et al. Approximation and Model Management in Aerodynamic Optimization With Variable-Fidelity Models [J]. Journal of Aircraft, 2001, 38(6): 1093-101.

[18] WANG H, ZHU X, DU Z. Aerodynamic optimization for low pressure turbine exhaust hood using Kriging surrogate model ☆ [J]. International Communications in Heat & Mass Transfer, 2010, 37(8): 998-1003.

[19] ZHANG M, GOU W, LI L, et al. Multidisciplinary design and optimization of the twin-web turbine disk [J]. Structural & Multidisciplinary Optimization, 2016, 53(5): 1-13.

[20] AIAA. Two-dimensional high-lift aerodynamic optimization using the continuous adjoint method [J]. Aiaa Paper, 2000,

[21] KOLLA M L, YOKOTA J W, LASSALINE J V, et al. Stowability Constraint Within a Two-Dimensional Aerodynamic Optimization Method [J]. Journal of Aircraft, 2012, 46(2): 696-9.

[22] DING S T, ZHANG G, YE L I, et al. Sensibility analysis of the air inlet position in the rotating cavity [J]. Journal of Aerospace Power, 2011, 26(8): 1681-7.

[23] XIE G, LIU J, LIGRANI P M, et al. Flow structure and heat transfer in a square passage with offset mid-truncated ribs [J]. International Journal of Heat & Mass Transfer, 2014, 71(4): 44–56.

[24] SHUITING L X X G T Z D. Flow and heat transfer in rotor-stator system with pre-swirl angle of 30° [J]. Journal of Beijing University of Aeronautics & Astronautics, 2007, 33(3): 290-3.

[25] FONSECA C M, FLEMING P J. An Overview of Evolutionary Algorithms in Multiobjective Optimization [J]. Evolutionary Computation, 1999, 3(1): 1--16.

IOP Conf. Series: Journal of Physics: Conf. Series **885** (2017) 012012 doi:10.1088/1742-6596/885/1/012012

Effect of deposition speed on the flatness and cylindricity of parts produced by three dimensional printing process.

Muhammad Fahad[1*], Mahmood Khalid[2], Muhammad Nauman[2], Maqsood Ahmed Khan[1]

[1]Department of Industrial and Manufacturing, NED University of Engineering and Technology, Karachi, Pakistan.
[2]College of Aviation Safety Management, Karachi, Pakistan.

*mfahad@neduet.edu.pk

Abstract. The idea of layer-by-layer deposition of materials to obtain three dimensional shapes, known as three dimensional printing, has gained much popularity during the last decade. Investigations related to understanding the effect of process parameters on the output of the accuracy of parts produced by three dimensional printing processes have been performed by various researchers. This study is also aimed at investigating the effect of deposition speed on the accuracy in terms of geometric dimensions and tolerancing such as flatness and cylindricity of the parts produced by open source three dimensional printers. The repeatability of the parts printed were also investigated.

1. Introduction

The concept of three dimensional printing (3DP) originated as rapid prototyping (RP) processes during mid-1980s. These processes were mainly aimed at producing a part in a tool-less manner by only using the computer-aided design (CAD) information of the part [1, 2]. The growth in technologies based on this concept of tool-less manufacture of products emerged to the level that the parts produced by these processes were successfully used in end-use applications and thus, were termed as additive manufacturing (AM) [3,4]. Society of Manufacturing Engineers (SME) and American Society for Testing and Materials (ASTM) define AM as "*A process of joining materials to make objects from 3D model data, usually layer upon layer, as opposed to subtractive manufacturing methodologies*" [5]. A variety of techniques have been established to that fall under AM such as [3]:

Stereolithography (SL): based on curing a liquid photopolymer into solid shape by using ultraviolet light (laser) beam.

Selective Laser Sintering (SLS): based on sintering of powdered materials (polymers, ceramics and metals) using a high power laser beam.

Fused Deposition Modelling (FDM): based on extruding beads of semi-molten thermoplastics via small sized nozzle.

The cost of all these proprietary machines were very high and thus presented the major hurdle in wide-spread use of these technologies. However, after the expiration of major patents during 2008, the development of low cost, open source machines by a large number of manufacturers have enabled these AM processes to be operated by an increased number of users. The term 3DP, therefore, become popular among the users of AM techniques and is used synonymously in conjunction with the AM.

Due to its relatively simple and low cost technique of building parts, the technology utilized in FDM became the most popular open source 3DP technology and a large number of manufacturers of this technology are available resulting in a relatively large user-base of this technology.
Comparison among various 3DP processes [6-8], as well as assessment of the geometric accuracy of different 3DP processes [9-12] have been reported by many researchers over the last 15-20 years. Recently, many researchers have presented reports related to accuracy of pars produced by open source 3DP processes utilizing FDM technology [12-15]. Various parameters including layer thickness, deposition speed, materials flow rate and raster width etc. have been varied to evaluate their impact on the accuracy of parts produced in the previously mentioned reports. Although, the accuracy of part shape in terms of dimensions (i.e. length, height, diameter etc) have been evaluated, but, none of these studies reported the effect of process parameters on the form of the object in terms of geometric dimensioning and tolerancing (GD&T) such as flatness, straightness, and cylindricity etc. Also, very little attention have been paid to the repeatability of the parts produced by open source 3DP. Therefore, in this study, the effect of deposition speed is investigated on the accuracy in terms of the dimensional measurement as well as GD&T of shape of objects produced by open source 3DP processes. The repeatability was also investigated by building 3 parts at each different speed.

2. Experimental

The geometric benchmark part designed by Fahad et al [16] was used and only the one third section of the part was built as indicated in Figure 1. The parts were built using prusa mendel i3 printer using a poly-lactic acid (PLA) filament. Both the machine and material were purchased from a Chinese supplier. Three different deposition speeds were used including 35, 45 and 55 mm/sec. Three replicates at each speed were printed to evaluate repeatability. The layer thickness was held fixed at 0.15 mm and the extruder temperature was 200°C for all builds. An infill density of 30% was used for all the parts. The parts were measured using a Wenzel X Orbit 55 coordinate measurement machine (CMM) and a 2 mm diameter, ruby probe was used.

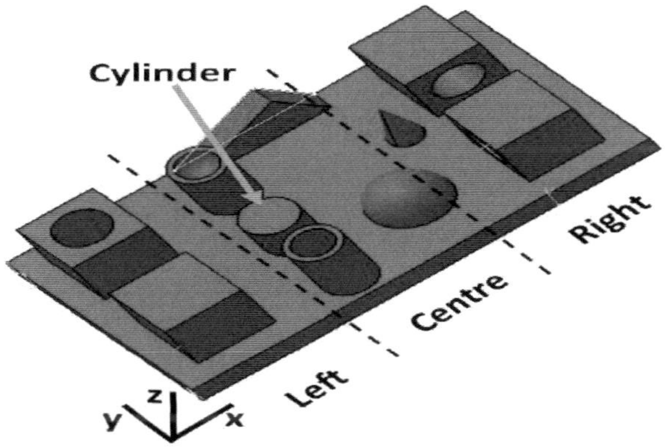

Figure 1. Benchmarking part used for measurements

3. Results and Discussion

3.1. Results

During the part build, slight curling of corners (i.e. left and right) was noticed therefore, it was decided to divide the base of part into three sections as left, centre and right (Figure 1) and five points within each section were measured to fit a plane and the flatness of each plane was recorded. The flatness values of each plane for all the parts are indicated in Figure 2. The solid cylinder (Figure 1) was also measured by taking a matrix of 4x4 points, that is, four points along the circumference of the circle at four different heights. The diameter and cylindricity of this solid cylinder for each part were recorded and the results obtained are indicated in Figures 3.

3.2. Discussion

As shown in Figure 2, it is very clear that the part base flatness deteriorates at higher speeds. Similar effect of speed on the accuracy of parts produced by open source 3DP process [14], however, for the purpose of comparison of flatness values, referenced can be made to a similar study conducted on sintering based 3DP processes [8]. This can be noted that the flatness values obtained are deteriorating with speed, however, the worst values obtained (i.e. near 0.4 for right section at 45 mm/sec, Figure 2.b) are better than those obtained for commercially established process (~0.7) [8].

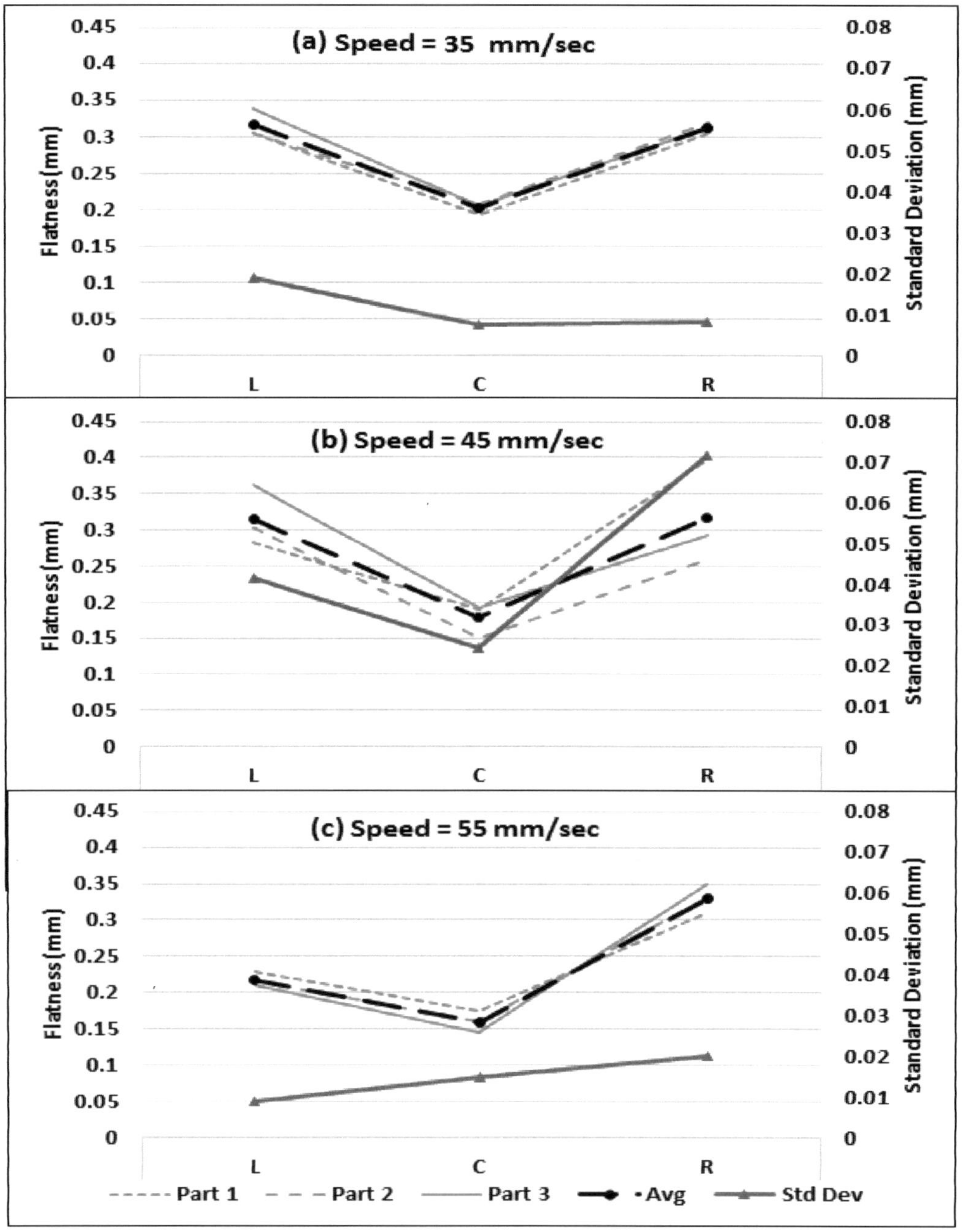

Figure 2. Flatness of left (L), centre (C) and right (R) sections of base at the three different speeds

It can also be noted for every speed, that the centre section has better flatness than the left and the right sections. This indicates the curling effect at the corner surfaces of the flat base for the part. The reason behind this flat base could be the same as reported by other researchers [8,17], that is, the thermal gradient. Due to the difference in temperature of the layer being deposited (hot) and the layer already deposited (cold), the hotter layer, during shrinking, pulls the previously deposited layer which had already shrink to attain its original dimension. This pulling effect causes the curling effect and thus the flatness at the centre is better than at the two corners. This can also be noted that not only the flatness values are better at lower speeds, but the repeatability is also better at slower speeds, this is indicated by a low standard deviation at 35 mm/sec (Figure 2.a) compared with those at 45 and 55 mm/sec. Similar trends were observed for the cylinder diameter and cylindricity value as indicated in Figure 3. The actual (designed) diameter of cylinder was 10 mm and it can be noted that at 35 mm/sec, the average value of cylinder dia was consistently near 9.8 mm (error of 0.2 mm) whereas at the highest speed, the average value of cylinder dia was near 9.85 m (error of 0.15 mm) but the variation among the three parts were high compared to those obtained for 35 mm/sec. Both the cylindricity values and the standard deviation in cylindricity values were better at 35 mm/sec compared with the two high speeds (i.e. 45 and 55 mm/sec). This, therefore, indicates that the slower speeds provide better accuracy in terms of geometric dimensioning and tolerancing and the repeatability of the parts is also better at slower speeds.

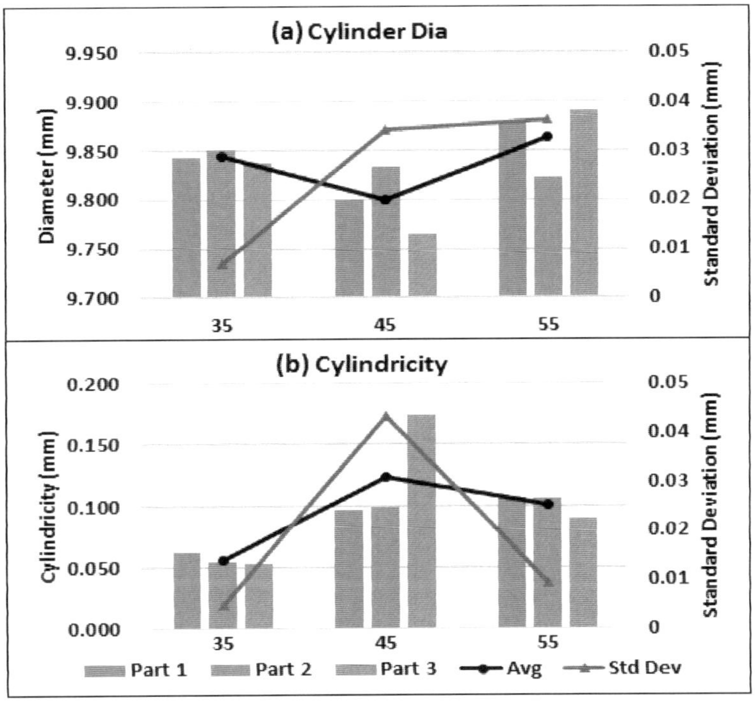

Figure 3. Values of (a) cylinder diameter and (b) cylindricity at different speeds

4. Conclusion

The accuracy and repeatability of parts produced by open source 3DP process was studied and the effect of variation in deposition speed was investigated. It was found that better geometric dimensions and tolerances were obtained at slower speeds. The obtained GD&Ts at slower speed also showed lower standard deviations compared with those at higher speeds indicating better repeatability at slower speeds.

References

[1] Dickens P. M., "Rapid Prototyping – The Ultimate in Automation?", *Assembly Automation*, 14(2), (1994), 10-13.

[2] Hopkinson N., Dickens P. M., "Rapid prototyping for direct manufacture", *Rapid Prototyping Journal*, 7(4), (2001), 197-202

[3] Fahad M., "Additive Manufacturing (Chapter 17)" Materials Processing for Engineering Manufacture (Ed. Z. Huda), Foundations of Materials Science and Engineering, 93, 2017, 357-378

[4] Perween S., Fahad M., Khan M. A., "Trends and Future Perspectives in Additive Manufacturing", Proceeding of Advanced Materials and Process Engineering (AMPE) Conference, 14-15th December, 2015, Karachi, Pakistan, 4-59

[5] A. Standard, F2792, Standard Terminology for Additive Manufacturing Technologies, *ASTM International*, West Conshohocken, PA, USA, 2012

[6] Ippolito R., Iuliano L., Gatto A., "Benchmarking of Rapid Prototyping Techniques in Terms of Dimensional Accuracy and Surface Finish", CIRP Annals - Manufacturing Technology, 4(1), 1995, 157-160

[7] Shellabear M., "Benchmark study of accuracy and surface quality in RP models", RAPTEC, Task 4.2, Report 2, 1999

[8] Fahad M., Hopkinson N., "Evaluation and comparison of geometrical accuracy of parts produced by sintering-based additive manufacturing processes", International Journal of Advanced Manufacturing Technology, 2016, doi:10.1007/s00170-016-9036-z

[9] Fahad M., Hopkinson N., "Evaluation of Parts Produced by a Novel Additive Manufacturing Process", Applied Mechanics and Materials, 315, 2013, 63-67

[10] Dimitrov D., van Wijck W., Schreve K., de bear N., "Investigating the achievable accuracy of three dimensional printing", Rapid Prototyping Journal, 12(1), 2006, 42–52

[11] Akande S. O., Dalgarno K. W., Munguia J., Pallari J., "Assessment of tests for use in process and quality control systems for selective laser sintering of polyamide powders", Journal of Materials Processing Technology, 229, 2016, 549-561

[12] Lanzotti A., Giudice D. M. D., Lepore A., Staiano G., Martorelli M., "On the Geometric Accuracy of RepRap Open-Source Three-Dimensional Printer", Journal of Mechanical Design, 137(10), 2015, 101703 (08)

[13] Cruz Sanchez, F. A., Boudaoud, H., Muller, L., & Camargo, M., "Towards a standard experimental protocol for open source additive manufacturing", Virtual and Physical Prototyping, 9(3), 2014, 151-167

[14] Lanzotti, A., Martorelli, M., & Staiano, G., "Understanding Process Parameter Effects of RepRap Open-Source Three-Dimensional Printers Through a Design of Experiments Approach", Journal of Manufacturing Science and Engineering, 137(1), 2015, 011017

[15] Lieneke, T., Adam, G., Leuders, S., Knoop, F., Josupeit, S., Delfs, P., Funke, N. and Zimmer, D., "Systematical determination of tolerances for additive manufacturing by measuring linear dimensions", 26th Annual International Solid Freeform Fabrication Symposium, Austin, Texas, USA, 2015, 371-384

[16] Fahad M., Hopkinson N., "A new benchmarking part for evaluating the accuracy and repeatability of Additive Manufacturing (AM) processes", Proceeding of 2nd International Conference on Mechanical, Production and Automobile Engineering (ICMPAE 2012), Singapore, 234-238

[17] Lück T, Baumann F., Keller B., Wiedemann B., "Material research and development for RP techniques at the IKP", Proceeding of 3rd European Conference on Rapid Prototyping and Manufacturing, Nottingham, UK: 1994, 309-325

IOP Conf. Series: Journal of Physics: Conf. Series **885** (2017) 012013 doi:10.1088/1742-6596/885/1/012013

A methodology to determine the level of automation to improve the production process and reduce the ergonomics index

Alejandro Chan-Amaya[1, a], María Elena Anaya-Pérez[2, b] Víctor Hugo Benítez-Baltazar[3, c]

Universidad de Sonora, Blvd. Luis Encinas y Rosales Hermosillo, Sonora México. C.P. 83000

[a]alejandrochan12@gmail.com, [b]elena.anaya@unison.mx, [c]vbenitez@industrial.uson.mx

Abstract. Companies are constantly looking for improvements in productivity to increase their competitiveness. The use of automation technologies is a tool that have been proven to be effective to achieve this. There are companies that are not familiar with the process to acquire automation technologies, therefore, they abstain from investments and thereby miss the opportunity to take advantage of it. The present document proposes a methodology to determine the level of automation appropriate for the production process and thus minimize automation and improve production taking in consideration the ergonomics factor.

1. Introduction

The rapid mechanization and automation development of production process during the las century, was largely driven by human desire to reduce the proportion of manpower and thus improve the productivity of industrial facilities. The industry aims to find approaches that allow the production of high quality products at low cost [1].

Companies are constantly looking for improvements in productivity to increase their competitiveness. The use of automation technology is a tool that has been proven to achieve this, but it does not ensure advantageous results. One of the keys for a successful investment and implementation of automated equipment is the choice of the right technology, the type and level of automation that best suits the company, its needs, objectives and prerequisites. There are companies that are not familiar with this process, therefore, they abstain from investments and thereby miss the opportunity to take advantage of existing technology [2].

In the 80's there was an ambition to create factories that were totally automated, nowadays most of the automatic systems within the manufacturing are semiautomatic, consisting of the combination of automatic and manual tasks [3]. Research has demonstrated the importance of integrating human beings and technology into automated manufacturing to support sustainable and robust manufacturing systems. The relationship between humans and technology can be seen as activities that are performed entirely manually to fully automate, sharing tasks between human and technology. This concept is called levels of automation (LoA) [4].

2. Levels of automation

Automation does not necessarily refer to modernization or technological innovation, For example, updating a computer with a powerfull system does not necessarily constitute automation, therefore, automation can be definded as a device or system that accomplishes (partially or fully) a function that was previously could be carried out by a human operator [5].

Frequently it is said that the progression from manual operations to fully automated operations is done in a single step, that is, when operators are replaced by modern robots or machines. However, this is not entirely true. Manual operations are defined as work done by hand and not by machine, and machines, in the same way are defined as instruments designed to transmit or modify the application of energy, force or movement. The manual term can then be defined as the work done without any tool or support [3]. This implies that automation is not all or none, but can vary across a continuum of levels, from the lowest level of full manual throughput to the highest level of complete automation [5]. Therefore, giving the tools to the user or other support to achieve the task, can be seen as increasing the level of automation [3]. Table 1 shows the different levels of automation. LoA, mechanical and information levels can be interpreted as a 7x7 matrix with 49 possible combinations. The 49 possible solutions are then limited to a relative maximum and minimum level and create a framework of possible solutions (SoPI) based on company goals, ergonomics, environmental issues, investments, among others [6].

Table1. Scale automation levels for mechanical and computerized tasks

LoA	Mechanical and Equipment Level of Automation (Mechanical LoA)	Information and Control Level of Automation (Information LoA)
1	Totally manual	Totally manual
2	Static hand tool	Decision giving
3	Flexible hand tool	Teaching
4	Automated hand tool	Questioning
5	Static machine/workstation	Supervision
6	Flexible machine/workstation	Intervene
7	Totally automatic	Totally automated

3. Case study

In areas of manual operations workers commonly make mistakes in their activities for different reasons, among them: distractions, fatigue and repetitive activities. Because of the above; has a direct impact on production and ergonomics, since in many cases the target is not achieve, thus raising the costs of manufacturing. The industrialization and process automation, especially in the industry of electric harnesses for the automotive market, must be with a very high level of quality and for this it is necessary an exhaustive supervision of each step to be realized during the manufacture.

The present work will be carried out in an automotive industry located in Mexico, the company is dedicated to carry out several processes, and one of them is the tape process of harnesses for automobiles, the process is showed in figure 1. The company seeks to increase its productivity and improve the ergonomic index through making use of technology.

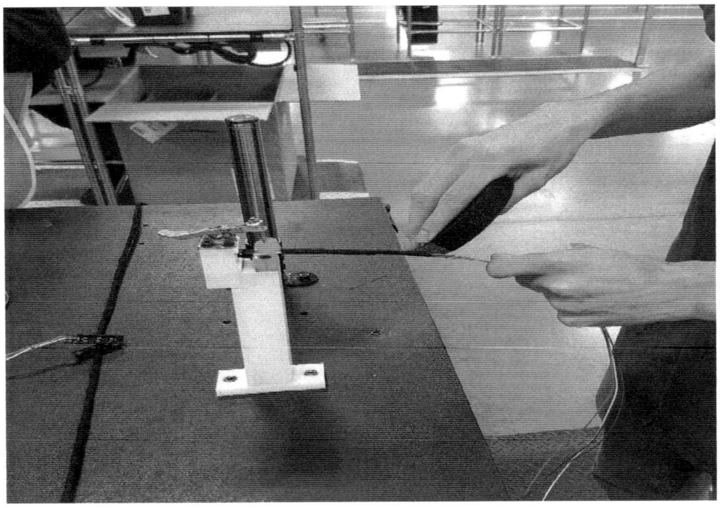

Figure 1. Tape process harness.

It is proposed a methodology that allows to make the selection of the appropriate level of automation which will lead to the result expected by the company, this methodology is based on a previous work developed by the authors Fasth et al. [7] consisting in 4 steps to develop, which is presented in figure 2.

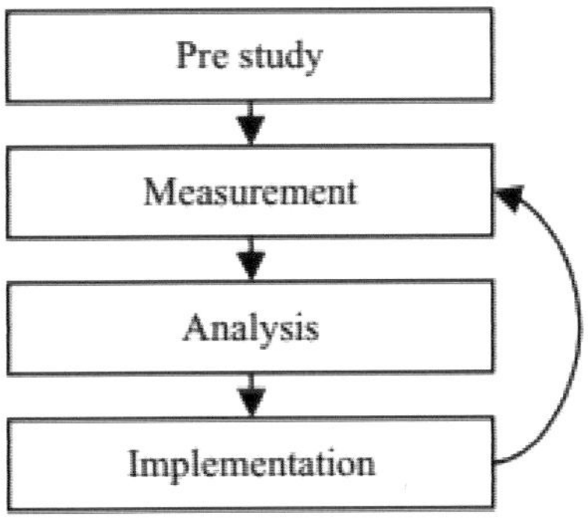

Figure 2. Phases in the measurement methodology [7].

3.1. Pre study
This step consists in defining in how it is the current situation of the process to study and to know the characteristics with which the operation is currently being carried out, the following is a list of the tasks that have to be performed in order to complete the current phase:

- Select process to study
- Walk the process and know it
- Perform Value Stream Mapping (VSM) to identify flows and parameters

3.2. Measurement
This phase aims to disaggregate each of the activities performed by the operators in each of the workstation stations, as well as to measure the level of automation in the process. The activities to be performed in this phase are shown:

- Ergonomic analysis
- Measurement of automation level
- Document results

3.3. Analysis

This phase of the methodology aims to determine the minimum and maximum level for each of the operations of the work center, with them design different proposals with the improvements that are intended to perform and finally analyze these possible improvements with respect to the parameters that were obtained in phase one, the tasks to be performed are the following:

- Conduct a workshop to determine the minimum and maximum level of automation
- Design of proposals
- Analysis of proposals

3.4. Implementation

The objective of the last phase is to visualize the different proposals that resulted in the previous phase based on the objectives and desire of the company, once selected the proposal will proceed to design the new work center (electrical, mechanical, Location of stations). In this phase, it is intended to make the implementation once the design is finished and to follow up to see the effect that was obtained after the implementation, it is worth mentioning that in this phase it is possible to return to phase two if necessary. This phase is divided into 4 tasks:

- View suggestions
- Station layout
- Implementation
- Tracing

4. Partial results

This chapter presents the partial results of the methodology proposed in the previous section, describing the results to date.

In order to do any work on this process, first step is to understand how it works, this is the goal that has this stage, in other words, information about the number of operators for that work center, the flow of the Parts within the process, the cycle times of each operation.

As a result of this stage was a lay out of the work center shown in figure 3, it can be observed the current design of the work center the location of the operators and the flow of the product and the VSM showing on figure 4.

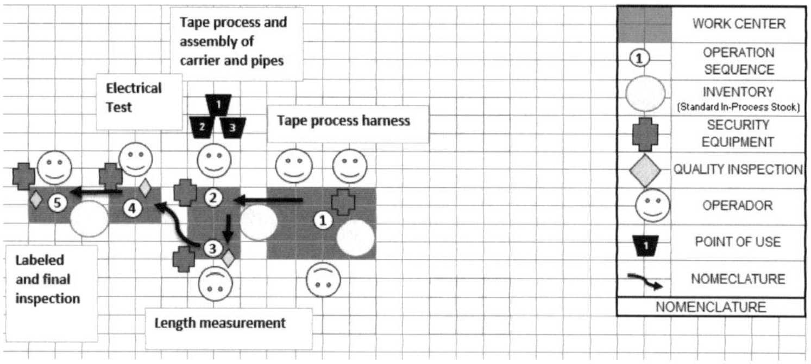

Figure 3. Plan of the work center under study (own design using company format).

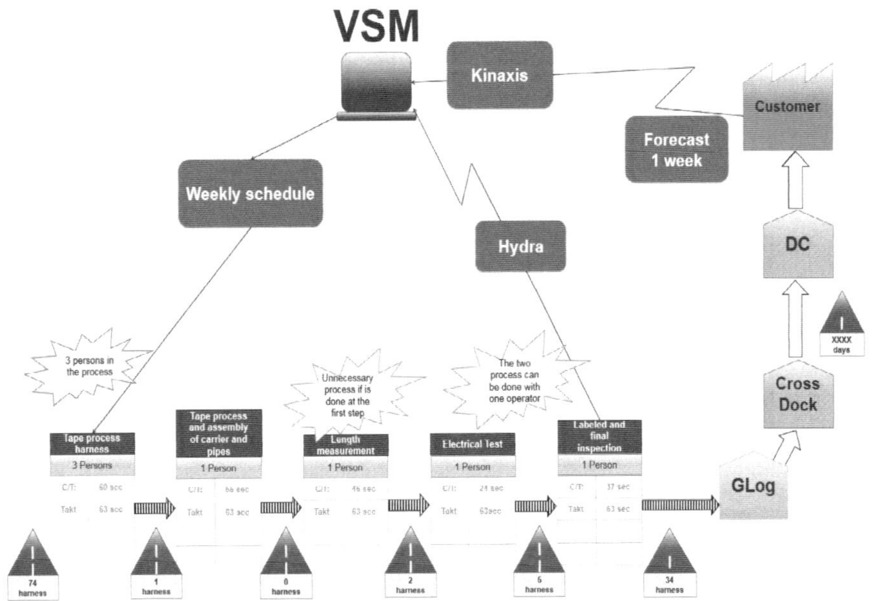

Figure 4. VSM of work center on study (own design using company format).

Developing a VSM, several possible improvements were found, and two of them are intended to be carried out with this project, in both cases reduce the number of operators and consequently the ergonomic index.
Once the process is known and the necessary parameters are obtained, the ergonomic study is carried out (this study was carried out by the Department of Safety and Hygiene), the results are showing on figure 5, where it is say the risk of performing is moderate. One of the company's goals are to improve this index.

Section I Scores	Neck	Shoulder	Back	Arm & Elbow	Wrists	Hands	Legs	Static Postures
	15	5	0	0	10	10	0	5

Risk Ranking		
	Greater than 85 = High 45 to 84 = Moderate 44 or Less = Low	45

Figure 5. Results of the ergonomic study (company format)

Currently, we are working on phase two where the measurements of automation level is carried it out and at the same time, we are looking for different ways to simulate the resulting future solutions.

5. Conclusion

With the implementation to a plant level of automatic systems in the different processes, the company can benefit not only production, but also performance, and product quality, as well as ergonomics. In order to obtain these desired benefits, it is necessary to determine and analyze the level of automation that the process requires for a better result and to achieve minimization of automation costs.
So that the above is possible to know the current situation of the company in terms of flow, times and automation, in order to know the areas of opportunity that has, in this way, make proposals guided by the needs and objectives of the company.

Acknowledgments
This research was supported/partially supported by University of Sonora.

We thank Rodriguez Hector for his assistance in giving the opportunity to develop this project inside the facilities.

References

[1] Aristova, N.I., 2014. Controlling Automation Level in Robot Production. Perfect and Stochastic Schemes. , 75(9), pp.1637–1645.
[2] Friedler, N., Salonen, A. & Johansson, C., 2013. The automation equipment acquisition process – experienced users' perspective
[3] Frohm, J., 2008. Levels of Automation in Manufacturing. , 1(28), pp.1–28.
[4] Lindström, V. & Winroth, M., 2010. Aligning manufacturing strategy and levels of automation: A case study. Journal of Engineering and Technology Management - *JET-M*, 27(3–4), pp.148–159
[5] Parasuraman, R., Sheridan, T. & Wickens, C., 2000. A Model for Types and Levels of Human Interaction with Automation
[6] Fasth, A. et al., 2009. SIMTER - A Joint Simulation Tool for Production Development
[7] Frohm, J., 2008. Levels of Automation in Manufacturing. , 1(28), pp.1–28.

IOP Conf. Series: Journal of Physics: Conf. Series **885** (2017) 012014 doi:10.1088/1742-6596/885/1/012014

Reduction of defective products by empiric analysis: the production line case in a factory of automotive parts.

Ramses Lugo-Telles[a], María de los Ángeles Navarrete[b]

Universidad de Sonora, Blvd. Luis Encinas y Rosales Hermosillo, Sonora México. C.P. 83000

ramses.lugo@hotmail.com[a,] mnavarrete@industrial.uson.mx[b]

Abstract. As all massive line production, in the production line under study, exist the possibility of produce defective products, so like all companies based in continuous improvement, it wants to know all the facts about these defective products, for example, how many are they? in which areas do they arise? Why do they arise? And among other questions, in order to suggest and implement some solution alternatives

1. Introduction

This paper was made in a manufacturing company of harness for the automotive production, located in Hermosillo, Sonora, Mexico. Through this study, we will see how some data of defective products can lead to the root cause. To clarify, the data provided in this paper, was generated during 28 weekends, the 4 lately weeks of 2015 and the first weeks of 2016.

On the other hand, an empirical analysis will help us to find the problematic root cause, because empirical analysis is a scientific model of investigation based on observation and experimentation [1]. Moreover, a defective product is when it does not meet with the standard characteristics established by the company [2], and a defective product is considerate as the main kind of west [3]. In the company under study the defects can be classified by 44 types, however, there are someone that are more important than others, for example, the three main defects are: Lock open, exposed filaments and no hipot point. Figure one shows a graphic of defects, where it can be observe that the defects mentioned before are the most important.

Through figure one we can observe that the defect named "Lock open" it's the one that generate more defects, however, we can know how the defect is named, but, do we know what is it? This is fundamental to propose solution alternatives.

IOP Conf. Series: Journal of Physics: Conf. Series **885** (2017) 012014 doi:10.1088/1742-6596/885/1/012014

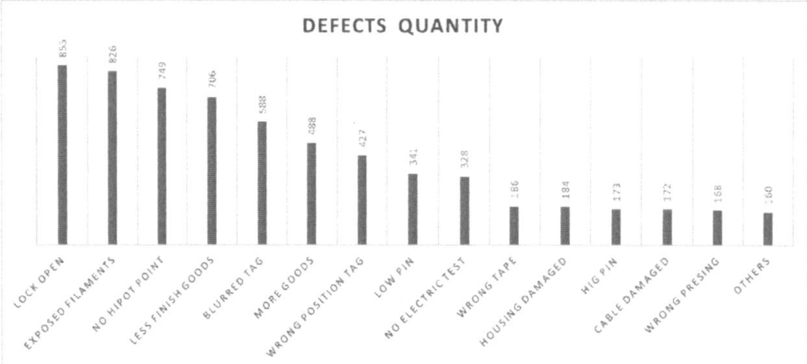

Figure 1. Defect Types (Own elaboration, 2016).

2. Backgrounds

To understand what is "Lock open" is necessary mention the main components of a harness. A harness is composed of 4 main components: The cable, sub-assembly, housings and locks. For a harness function in the correct way this component must work as the following way: The sub-assembly is adhered to the cable, one time adhered, the housing is assembled in the sub-assembly and finally the lock is stuck in the housing to avoid that the sub-assembly get separate from the housing (Go to figure two).

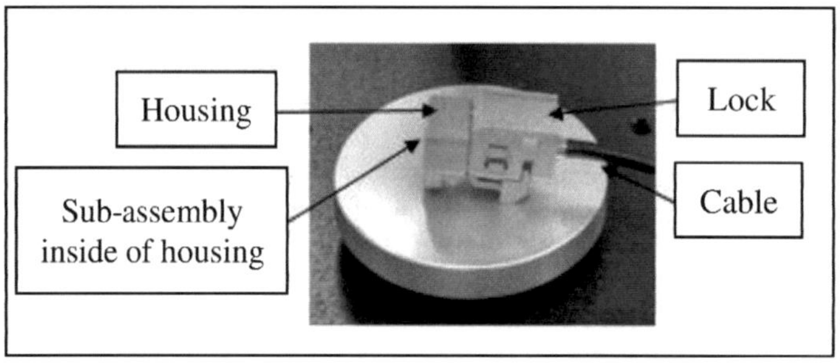

Figure 2. Harness components (Own elaboration, 2016).

When a lock does not get down until the housing is safe, the "Lock open" appears (go to figure three).

Figure 3. Lock open defect (Own elaboration, 2016).

One time defined what lock open is, it identified in what product (part number) it comes frequently this kind of defect. The part number with more "lock open" are show in the figure 4.

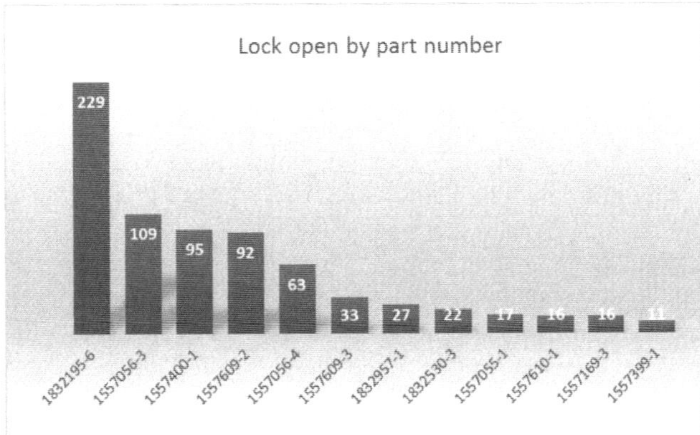

Figure 4. Lock open by part number (Own elaboration, 2016).

The figure above show us that the part number 1832195-5 with 229 defects, it's the one that more lock open generate. To understand better the defects nature of this product go to the figure 5(a), which is a pie chart that show the kind of defects and their percentage.

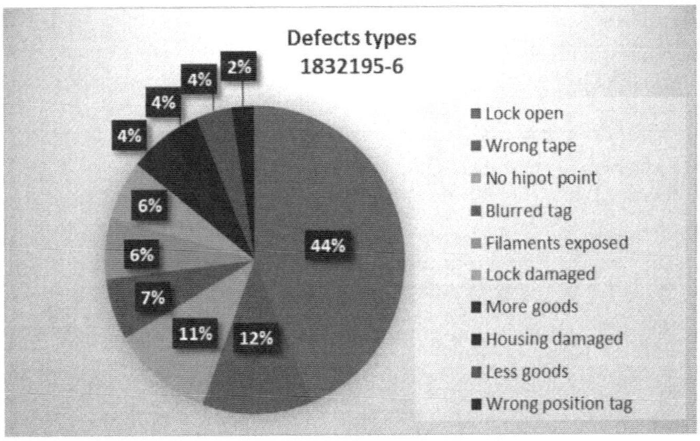

(a)

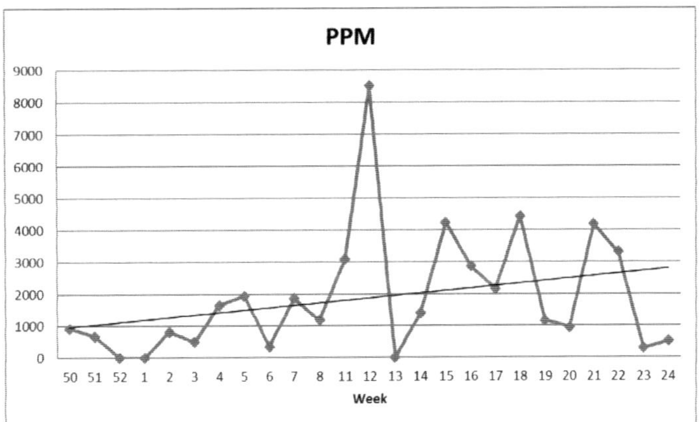

(b)

Figure 5. Kinds of defects of the 1832195-6 part number (Own elaboration, 2016).

In the figure 5(b) we can realize that of 229 defects shown in the part number under study, the lock open represents the 44% whence, it's considerate the principal factor to cause defects in this part number. This numeric data made us research closely about this defect, particularly in this part number. However, despite the analysis done until this moment, some questions arise: The lock open is already something of the past? Why this defect happens? And, is it a defect that will be more frequent?

3. Analysis and discussion

In order to initiating the analysis, it has been created a time chart, which it shows Parts Per million (PPM) of lock open defect, starting from de week number 50 of 2015 year, and ending at the week number 24 of the 2016 year, moreover, to the time chart it has been added a trend line, that displays the growing data.

It is worth to mentioning, the missing weeks in the chart above, it´s because there was not an inspection of that part number, because there was not production that days. During the problematic analysis, it questioned the way that the operators did the assemble between the housing and the lock. To the operator can do this assemble (in this part number in specific), is necessary that the operator wear bandage in their fingers. However, to produce others part numbers this bandages are not necessary.
In a quick talk with the operator that produce this part number (1832195-6), all of them agreed that the lock for this part number is more difficult to assemble than others, moreover, because of this kind of lock that it came a 90 grades clip, made the cable bend 90 grades (go to figure 6).

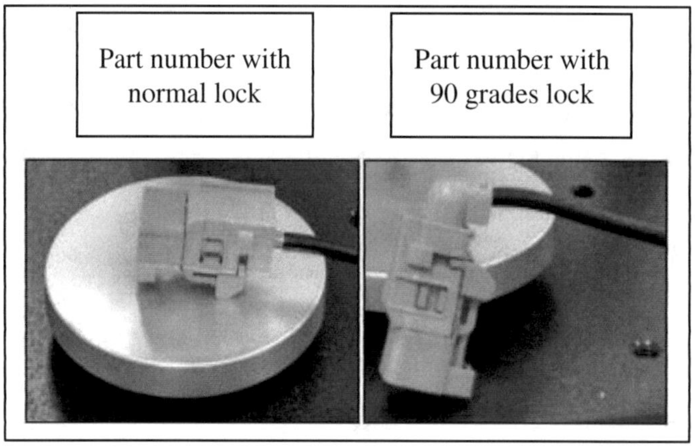

Figure 6. Normal Lock vs 90 grades lock (Own elaboration, 2016).

The operators supposed that for this kind of assembly they need more force in their finger. It made an experiment in the own company lab of metrology. The experiment consisted in realize three times the assemble with a force gauge. The first time, it was made with no cable and, only the housing and a normal lock. The second time was made with cable, normal housing and the lock, and finally, the last experiment, was made normally: cable, housing and the 90 grades lock. The results of these experiments are show in the table 1.

Table1. Experiment results.

Assembled with:	LbF (Pounds strennght)	Comments
Housing and lock	8.06	Any part numbers go with this assemble
Housing, lock and cable	13.28	The most of part numbers (except the under study product) the assemble is like this.
Housing, cable and 90 grades lock	54.28	In this way is how the part number under study is manufactured.

4. Conclusions

Based in the figure 5 we can say that the lock open quantity in the part number 1832195-6, it will rise according with the trend line. Hence the importance of attack this kind of defect. In other words, the lock open is going up, moreover, is one of the most common of all kind of defects in the production lines in the company under study.

According with the table 1, the difficulty of this kind of assembly is because of 90 grades lock. The same table shows a big gap between the LbF used in the assembly with a normal lock and the other one assembled with a 90 grades lock, therefore in this part of the analysis we can realize that the 90 grades lock made the assembly more difficult, however, this component (90 grades lock) it can't be removed of this part number, because is a request of the client.

To avoid this type of defect, the authors decided to implement a fixture that helps with this assemble. This fixture is intended to be a semi-automatic, this in order that the operator just put the components into a fixture, and it active a sensor (or a button), and the fixture would do the task currently made by an operator. By the implementation of this fixture we will avoid the fingers weathering of the operators, and by this way we care the health of these. Furthermore, the quantity of open locks it would significantly reduce, because of the fixture would do this operation, so the process variation drops, according to the settings of the fixture, which are manipulated by the engineering team of the company.

Finally, we conclude that an empiric analysis can help to solve a problem from the root, or well, it can let find the problematic from an unfavourable situation, moreover, trough by the same analysis find the solution.

References

[1] Radrigan M., 2005, Metodología de la investigación. Nueva Lente edicion. Mexico.
[2] Ruiz, C. & Marín, I., 2006, Producto inseguro y producto defectuoso.
[3] Southworht, T., 2010, Muda, Muri Muri. Printing Lean, 15(8), pp. 32-33.

IOP Conf. Series: Journal of Physics: Conf. Series **885** (2017) 012015 doi:10.1088/1742-6596/885/1/012015

Synchronization of workshops, using facilities planning

Britel Zineb and Cherkaoui Abdelghani

EMISYS: Energetic, Mechanic& Industrial systems. Mohamed V University EMI Mohammadia School of Engineers, Ibn Sina Avenue B.P.765Agdal Rabat, Morocco.

britelzineb@gmail.com

Abstract. In this paper, we will present a methodology used for the synchronization of two workshops of a sheet metal department. These two workshops have a supplier-customer relationship. The aim of the study is to synchronise the two workshops as a step towards creating a better material flow, reduced inventory and achieving Just in time and lean production. To achieve this, we used a different set of techniques: SMED, Facilities planning...

1. Introduction

In the automotive industry, achieving a one piece flow and producing in just in time is essential for lean production. Ever since its origin in Japan at Toyota in the early 1970s, JIT has become a "comprehensive management philosophy" [1]. The benefits of adopting just in time in manufacturing have been acknowledged by many industrial firms, as stated by Monden:"If JIT is realized in the entire firm, then unnecessary inventories in the factory will be completely eliminated, making stores or warehouses unnecessary. The inventory carrying costs will be diminished and the ratio of capital turnover will be increased" [2].
Functioning in Pull systems and Just in time manufacturing is indeed known for eliminating waste and improving productivity, however their down side according to Teruyuki Minoura, former President of Toyota Motor Manufacturing in North America: " If some problem occurs in one-piece flow manufacturing then the whole production line stops. In this sense it is a very bad system of manufacturing. But when production stops everyone is forced to solve the problem immediately." [3]
Therefore, investing in one piece flow processes is a good start towards lean production that leads to improvement overtime. From this perspective, we conducted a study about the synchronization of two workshops of a sheet metal department: The assembly workshop with the metal fitting and finishing workshop at an automotive assembly firm. These two workshops have a supplier-customer relationship. The aim of the study is to synchronize the two workshops as a step towards creating a better material flow, reduced inventory and achieving Just in time and lean production.

2. Methodology

2.1. Production plan analysis

To achieve the synchronization of the two workshops, we first analyzed the production plan, that is considered as an unchangeable constraint due to the fact that it is based on all the processes constraints, moreover there are other processes for which a change in the production plan would prove to be costly (expensive changeover of tools and wasting of residual raw materials…). Since the two workshops

Work in a push fashion, we need to know for each finished product of the customer workshop, what pieces the supplier workshop needs to provide and how many tool changes to operate if we want to respect the production plan in a one piece flow model. This analysis helped better understand the production needs in terms of products specificities and tool changes.
Besides, the supplier workshop suffers from an important storage activity that leads to traffic congestion within the plant and a complicated physical flow of materials (transport and material handling between the workshops and the storage area, overcrowding of semi-finished products…), hence the will of the company to reduce it through the synchronization project. The main cause behind this over-stocking issue is due to the "over-production" of 450 s (7.5 hours) in advance by the supplier workshop. The production plan is daily transformed into lots instead of one piece at a time to minimize the number of tool changes. Not only does the current work in process stock contributes to minimizing changeovers of tools but it also serves as a buffer stock and thus preventing chain stops. Nevertheless, as a form of "Muda", inventory needs to be reduced if not eliminated.

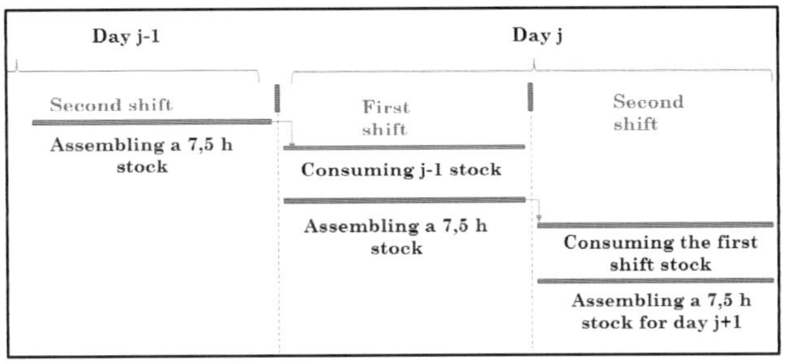

Figure 1. The Initial stock replenishment mode

During this study, we realized that the main constraint in the synchronization of the workshops was due to the important number of specificities in the finished goods produced, leading to an important number of changeovers (of tools, machinery and raw materials), that is avoided by the work-in-process inventory produced by the first workshop. However, in the synchronization context, all tool changes imposed by the production plan shall be taken into account. Consequently, we tried to simulate a single piece flow on an excel sheet using the current tool changing time that varies between 420s to 600s (7 to 10 minutes), according to the type of tools or machinery used for each change. We realized afterwards that given the actual tool changing time, it won't be possible to produce the minimum required quantity of finished goods per day, hence the need to reduce the time of tool changes.

2.2. A SMED Approach

The changeover of tools only happens for the first workshop, for the second one, only the raw materials are changed. In order to reduce the changeover time, we relied upon a SMED approach (Single Minute Exchange of Die). Following a detailed analysis of the different types of changeovers that occur for the first workshop (route sheets, flow and process analysis, operations analysis, Gantt diagrams…), it became clear that the transportation and storage of the tools and machines are the primary consuming time activities and thus we decided to see if we could remove those activities.
The table below sums up the main problems encountered and the proposed solutions:
Many of the proposed solutions following the SMED analysis have in common the change of the current disposition of the first production workshop as a mean to reduce changeover times. The idea here is to group all of the production means (tools and machinery) that are usually stored in another storage area in the same work cell where they are needed. A suitable facilities disposition would allow us to : avoid flow crossing, avoid congestion inside the workstation, store the tools and machinery on the ground and avoid their transportation, bring the storage area closer to the station or put all the tools inside the station. The storage problem (for tools and machinery) would be solved and our gains would be in terms of time and space. Consequently, the next issue to address is how can we organize those

production means in the space-limited work cell, without disrupting the flows, encumbering the cell and inconveniencing the operators?

Table 1. Summary of the SMED analysis findings

Operation	Responsible causes for increasing time	Parameters to study	Proposed solutions
Operation 2	Difficulty of access and clutter for the stacker	-WorkStation intern disposition -Difficult access to the station.	-Proposing a new implantation for the workstation -Standardization of changeover paths
Operations 3, 9	Heavy load to transport	Heavy load	-Consider alternative handling means -Avoid moving tools and machinery
Operation 4	Maneuvering a heavy load to an elevated position.	Storage in high places	Storage of tools and machinery on the ground
Operation 6	Recurring position error	Positioning errors	Using POKA YOKE
Operation 7	Free transport (transporting nothing)	Distance between the storage area and the workstation	-Moving the storage area closer to the workstation. -Putting all the tools and machinery inside the workstation.
Operation 8	The shape of the tool involved makes its grip more difficult	Shape and volume of the tools	Taking it into consideration

2.3. Facilities planning

We started by calculating our space needs and comparing them to the surfaces available in the work cells which makes our solution feasible in terms of space. Since we do not have an initial facility plan for the desired solution, we started by the study of the initial implementation through a flow diagram, it is usually done on a real map of the plant, and it shows the path taken by each piece and represents the total plant flow. This approach is useful for detecting flow crossing and backtracking, and gives an idea of the covered distances. Flow crossing can be responsible for congestion and security issues, whereas Backtracking refers to "material moving backward in the plant… Backtracking costs three times as much as flowing correctly" [4]; indeed parts should be routed to the "customer" service and not in the direction of the "supplier", hence the need to eliminate it.

In our quest to a better flow, we used different flow analysis techniques: String diagrams and multi-column process charts. These two methods rely upon the creation of "route sheets" according to the route sheet data requirements specifies in the work of [5] "Facilities planning".

- String diagrams: a representation where circles refer to equipment and lines to the flow, in case of backtracking the line goes beneath the circles. [4]
- Multi-column process chart: It is based on the calculation of the number of steps travelled. A step refers to the distance between the centres of two consecutive circles, meaning if we jump a circle we will have two steps. [4]

Using these techniques, we evaluated the efficiency of each proposed plan for a new and improved layout. Here below is an example of a multi-column process chart with the efficiency calculus.

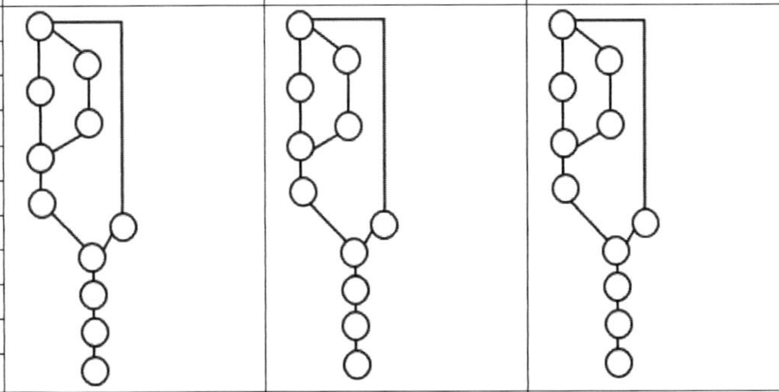

MULTI-COLUMN PROCESS CHART				
Part Numbers				
Tools	N 1	N2	N3	
R				
M1				
M2				
M3				
M4				
M5				
M6				
M7				
M8				
M9				
S				**Total**
Number of steps	18	18	18	**54**
Least steps	10	10	10	**30**

Figure 2. Example of a multi-column process chart

$$\text{Efficiency} = \text{Least steps} / \text{Number of steps} \quad (1)$$

For the first proposition, we found: Efficiency=30/54=55,55%. We repeated this steps for different layout proposals, and thus moving from an improved efficiency to another, until we reached the chosen layout; The improvements adopted consisted in avoiding loops between workstations (getting two machines 1 and 2 out of the workstation but still very near) and the installation of rails on which the machines will slide during tool changes. The efficiency shifted from the initial 55,55% to 58.82%. Here below are the initial and the final layout; X, Y and Z refer to the empty place where the machine/tool will be slided to on the rails.

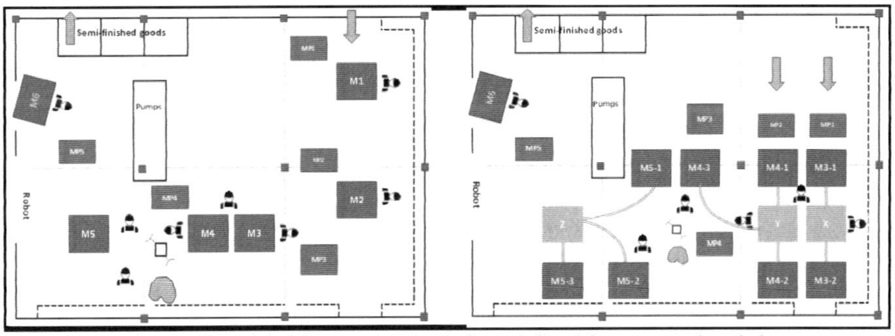

Figure 3. Initial layout at the left and final layout at the right

We used a time standard analysis MTM1 (Method Time Measurement 1) in our case, to estimate the new tool changing time that has become 4 minutes when using the improved facility design instead of (7 to 10 min before). It is mainly due to the possibility of changing all of tools/machines simultaneously instead of one by one as it used to be done before.

2.4. Synchronization scenarios

To make the synchronization of the two workshops possible, we studied four scenarios. Each scenario presents an idea for synchronizing through a special procedure. All these procedures come together in the act of delivering the right piece to the right place in the right sequence at the right time from the first workshop to the second. However some scenarios have drawbacks that cannot be overlooked. The main differences between the four scenarios are either to produce per batches or not, and how many batches per time slot should we consider. We used Gantt diagrams on Excel (that represents activities

in scales of hours, minutes and seconds unlike their classical use in project management where they rely on day scales) to simulate the production on a day where we have the highest amount of tool changes (for the last three years), we could access this data from the company history and applied our scenarios for that specific day.
The chosen scenario consists, for each time slot (of one hour), in producing pieces in batches according to their specificities. Thus, minimizing the number of tool changes per hour, while achieving the workshops synchronization; for each time slot, we deliver the right piece in the right amount with the right specificities and at the right workshop. We also noticed that this scenario allows us to produce more parts (at least 5 more per day) which is a good starting step if the plant ever wants to rise its current production capacity.
The remaining problem appears to be in the sequencing of pieces, because even though we produce the right products for each time slot they are not produced nor delivered in the right order and this is why we proposed to develop a VB.Net application that helps the operators in placing the right product in the right place of the metal containers that are transported through forklifts between the two workshops. The metal containers used have metal separators for each product to be placed which makes our VB.Net application very useful, the last operator to receive the piece assembled has simply to put it in its right place of the metal container as shown in the computer application in front of him.

2.5. *Further considerations*
With the transition to synchronisation, producing or assembling defective parts and rebus require retouching or outright remake, which consumes time and can lead to accumulating delays or chain stops, hence the idea of having a safety stock next to the assembly stations and in case of a defect we replace it right away. Thus, we calculated the quantity of pieces to keep as our safety stock to replace any defective product, we relied on the defects history to determine which and how many pieces to store to avoid delays in current timeslots. In addition, we will just have to make sure to renew our safety stock every week to avoid oxidation issues.
We also decided to reorganize the workload among operators, and to assess the synchronisation influence on logistics activities according to the assumption of the new operating mode, starting with forklift drivers that transport the metal containers back and forth between the two workshops, we used the MTM3 (Method Time Measurement 3) method to evaluate their workload before and after the synchronisation of the workshops, and reorganize their activities to avoid imbalances.
Finally we studied the project profitability, and the risks of operational failures through a failure mode effects analysis (FMEA). It is essential, when dealing with similar projects to summarize the results in a clear and concise procedure and to communicate it with all those affected by the changes undertaken. A communication-based approach involving all employees can help overcome problems of resistance to change that may result from the synchronization project.

3. Conclusion
In this study, we explored the synchronisation of two workshops in an industrial context. We started with an analysis of the production needs for a synchronized process between the two workshops with an emphasis on the frequency of tool changes. Especially since the synchronization requires to perform all of the tool changes imposed by the plant production plan. Hence the need to reduce their time. The idea that we came up with is considering a new flexible workstation design that will enable the achievement of tool changes in less time. We also proposed several synchronization scenarios and carried a careful study to choose the most advantageous one. Finally, we calculated the safety stock to use and focused on the problems that may arise with the synchronization through a risk study. In conclusion, all of the steps mentioned represent a step by step guide to exploit this factory's experience in the synchronisation of workshops. This experience can benefit other companies and can even be applied in other industries.

References

[1] T. Cheng and S. Podolsky, Just in time manufacturing, an introduction, 2e ed., Chapman & Hall, 1993.
[2] Y. Monden, Toyota production system; An integrated approach to just in time, 4e ed., CRC Press, 2012.
[3] D. J. K. Liker, Toyota Way: 14 Management Principles from the World's Greatest Manufacturer., McGraw-Hill Professional, 2004.
[4] F. E.Meyers and M. P. Stephens, Manufacturing facilities design and material handling, 3e ed., 2005.
[5] J. A. Tompskin, J. A.White, Y. A. Bozer and J. Tanchoco, Facilities planning, 3e ed., 2003.

IOP Conf. Series: Journal of Physics: Conf. Series **885** (2017) 012016 doi:10.1088/1742-6596/885/1/012016

Simulation based optimized beam velocity in additive manufacturing

Frédéric Vignat[a,*], Nicolas Béraud[b,*], François Villeneuve[a,*]

[a]Univ. Grenoble Alpes, CNRS, G-SCOP, F-38000 Grenoble, France
[b]DP Research Institute, F-34000 Montpellier, France

Frederic.Vignat@grenoble-inp.fr

Abstract. Manufacturing good parts with additive technologies rely on melt pool dimension and temperature and are controlled by manufacturing strategies often decided on machine side. Strategies are built on beam path and variable energy input. Beam path are often a mix of contour and hatching strategies filling the contours at each slice. Energy input depend on beam intensity and speed and is determined from simple thermal models to control melt pool dimensions and temperature and ensure porosity free material. These models take into account variation in thermal environment such as overhanging surfaces or back and forth hatching path. However not all the situations are correctly handled and precision is limited. This paper proposes new method to determine energy input from full built chamber 3D thermal simulation. Using the results of the simulation, energy is modified to keep melt pool temperature in a predetermined range. The paper present first an experimental method to determine the optimal range of temperature. In a second part the method to optimize the beam speed from the simulation results is presented. Finally, the optimized beam path is tested in the EBM machine and built part are compared with part built with ordinary beam path.

1. Introduction

The quality of metallic parts manufactured with additive technologies is highly related to melting strategies. Melting strategies means beam trajectories and characteristics such as speed, intensity and focus. These strategies are based on simple trajectories built on contour and hatching strategies filling the contours at each slice. Beam characteristics are determined from simple thermal models to control melt pool dimensions and temperature and ensure porosity free material. These models take into account variation in thermal environment such as overhanging surfaces or back and forth hatching path. However not all the situations are correctly handled and precision is limited. This paper proposes new method to determine energy input from full built chamber 3D thermal simulation. Using the results of the simulation, energy is modified to keep melt pool temperature in a predetermined range. The paper present first an experimental method to determine the optimal range of temperature. In a second part the method to optimize the beam speed from the simulation results is presented. Finally, the optimized beam path is tested in the EBM machine and built part are compared with part built with ordinary beam path.

2. Literature review

Numerous studies deal with the simulation of powder bed additive manufacturing process. [1–3] proposes thermal simulation of the electron beam additive manufacturing process based on a finite element model. [4–7] propose it for laser additive manufacturing. These models are then used for predicting the thermal stress and deformations or the size of the melt pool. [8] has used these models to predict the effect of support structure on thermal deformations and proposes new support strategies. [9] used this model to predict the size of a built strut in EBM and thus improve dimensional accuracy. [10,11] proposes a simulation of the process based on a Boltzmann lattice model where the scale of the study is the powder grain (50 μm). This study aims to better understand the EBM process on a microscopic scale and to make recommendations in order to better master the process. They also used their simulation to optimize the hatching strategy in EBM as presented in [12].

These models are effective to simulate millimeters of trajectory. However, the simulation of a complete manufacturing entity requires simulating several meters, or kilometers, of trajectory. The finite element simulation is suitable for the understanding of the process. However, a faster simulation would allow to increase the size of the simulated volumes and thus could be used to assist in the choice of the manufacturing parameters or their optimization. This is why a new simulation was developed based on different principles as proposed by [13].

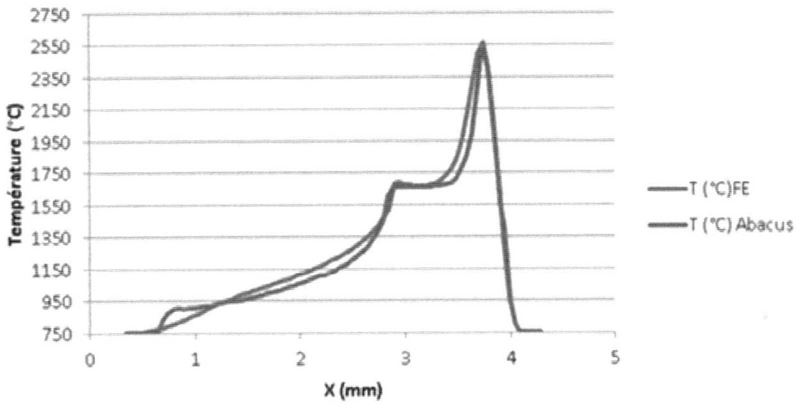

Figure 1. Temperature along a melted line, comparison between Abacus and Finite Element

The founding idea of abacus simulation [13] is to see the simulation of a complete case as the succession of standard cases. The standard cases are the heating and cooling of the material under a fixed electron beam. From these standard cases, graphs of evolutions of the temperature of the different points of the 3D space are generated. They store the variation of the temperature of the points around the beam center as a function of the beam parameters and the position of the point relative to the beam center. The displacement of the electron beam is modeled as an electron beam remaining fixed for a time step and then moving to the next point. At each time step, the evolution of the temperatures of the material is known thanks to the charts calculated with a fixed beam. The results obtained using Abacus simulation has been compared to results from finite element simulation as proposed by [1,2]. Figure 1 shows a temperature distribution along a line swept by the beam. In this case the beam has traveled from X=0 to X=4 mm and the picture is taken when the beam is ate X=4mm. This figure shows limited difference between FE considered has the reference and abacus that needs to prove its reliability. Comparison with experimental results would have been better but no data exists to make it as measuring temperature distribution in an EBM machine is still an unsolved issue. Based on comparison with FE results the abacus simulation has been considered has precise enough to develop simulation based melting strategy optimization.

3. Determination of the optimal range of temperature

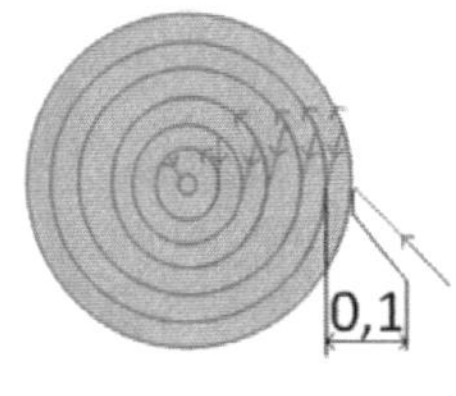

Figure 2. Sample disc melting trajectories

The abacus based simulation proposed by [13] provides the temperature distribution over the built part. This temperature must be controlled no only to limit thermal distortion and porosity. Thermal distortion is a consequence of overheating and porosity of under heating. It is thus necessary, to later create optimal beam path, to determine the optimal temperature range. In order to establish this link, a disc of radius 8 mm and thickness 5 mm is manufactured. The beam trajectory used is composed of 79 concentric circles as shown figure 2 and the melting parameters (speed, current, focus) are kept constant along the path. The built of the disc is simulated using the abacus based simulation and the temperature distribution along the disc radius is given figure 3. The circle becoming smaller at the center, the temperature reached at the center is higher than for the external zones.

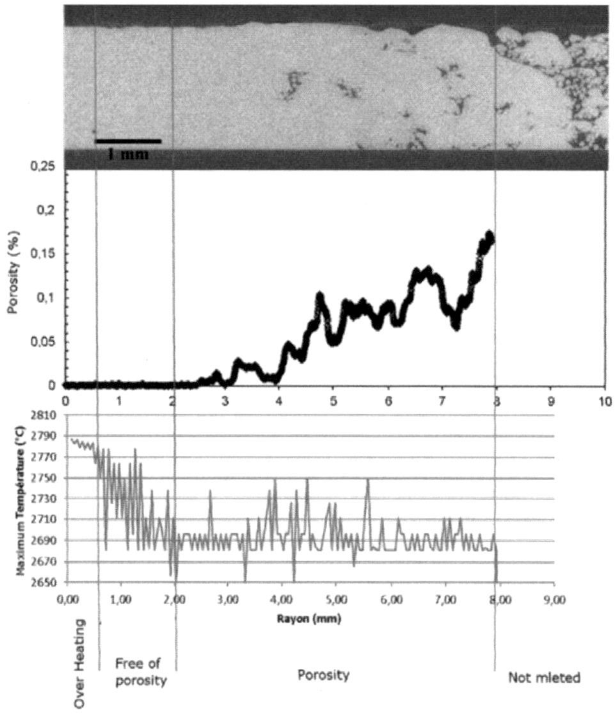

Figure 3. Temperature and porosity along disc radius

This part is then manufactured and a section is cut to be analyzed using tomography. Figure 3 shows a view of this section and the porosity measured along the radius from the tomography results. A correlation between the maximum temperature reached and the porosity rate can be observed and four areas can be distinguished:

- The unmelted zone where the temperature of the material did not exceeds 1665 ° C.

- The area where the material is molten but porous. It corresponds to temperatures between 1665 and 2700 ° C. However, there is no correlation between this temperature and the porosity rate, only the presence of porosities can be predicted.
- The area where the material is not porous. It corresponds to temperatures between 2700 and 2760 °C. Here, the material is correctly melted and the number of porosities is minimal.
- The central area where the temperature exceeds 2760 °C, the rate of porosities is minimal but a shape defect of the upper surface of the disk is observed. This is why this zone is considered overheated. The upper surface of the workpiece seems to be correlated to the excessive temperature reached.

Looking at these results, the optimal range of temperature is determined as being between 2700°C and 2760°C.

4. Simulation based optimal beam path

Most of the melting strategies are determined using very simple thermal models. Having a full 3D process simulation allows using simulation-based optimization method as proposed by [14] to optimize the different parameters of the fusion strategy. For the optimization of EBM melting strategies, the method is built on successive steps. First, an objective function is defined: in the current case study, all points of the last manufactured area have to reach a maximum temperature between 2700 and 2760 ° C. Then the optimization parameters are defined: for the EBM process beam trajectory and beam parameters along the trajectory are considered. Then for each time step:

- The melting strategy parameters are optimized based on the results of the previous time step simulation. The optimization function provides new parameters (beam current, beam focus and new beam position) for the current time step t to comply with the objective function. The optimization is based on thermal result from the previous time step (t-dt). It modify the energy input for the current time step to control the temperature range based on the principle of automatic control.
- The process is simulated with the new parameters and a new heat map for the current time step (t) is calculated. The temperature reached by the current beam center at the current center will be used for the next time step melting strategy parameters optimization.

This method makes it possible to define the melting parameters at time t (position and parameter of the beam) based on the temperatures reached at time step t-dt and the target temperatures set. The optimization function therefore acts as a servo function of the melting strategy over time depending controlling the temperatures reached. The trajectory followed during the simulation will thus respect the thermal criteria and can then be used for manufacturing.

One of the advantages of this method is the separation of manipulated objects:

- Simulation of the process.
- Optimization function: in the following example, a simple optimization function will be proposed but the search for the best optimization function is a scientific challenge as such. Indeed, the thermal phenomena involved are highly non-linear and multi-variable.
- Objectives: in the following example the quality criteria defined above have been used, however, if the objectives in terms of quality change, these could be modified without affecting either the simulation or the optimization function. For example, for certain applications such as heat dissipation, the presence of porosity in the material may be allowed. In other cases, a high geometric quality will be required but porosities will be tolerated. The target temperature will then be modified but this won't affect the simulation or optimization module.

Such an environment allows a great flexibility in terms of objective, optimization function and simulation.

5. Case study

Table1. Melting parameters

Focus Current	If	4
Initial speed (mm/s)	V	980
Beam voltage (kV)	U	60
Beam current (mA)	I	5
Layer thickness (µm)	h	50

For the case study the goal is to determine the optimal strategy to melt a prismatic part as represented figure 4. In the first approach, the trajectory is fixed as well as the beam current and beam focus. Only the beam speed along the path is modulated to control the temperature reached. The trajectory used is a simple hatching as represented figure 5 and the beam parameters are given table 1. The optimization function will increase the speed if the temperature gets too high and will come back to the initial speed if the temperature is too low. The proposed function varies the speed along a predefined path in order to meet the quality criterion. The objective is to stay in the optimal temperature range between 2700°C and 2760°C. For this, a target temperature *Ttarget* was set at 2730°C i.e. the middle of the optimal temperature range. The speed variation is then calculated based on the maximum temperature at the previous time step *Tmaxt-dt.* Three cases are distinguished:

- If the temperature is below the threshold temperature set at 96% target temperature, i.e. 2620°C, the speed is set to the initial speed (*InitialSpeed*).
- If the temperature is between the threshold temperature and the target temperature, the speed is increased and a new speed (*Speed*) is calculated using equation (1).
- If the temperature is above the target temperature, the speed is increased such that the next simulated point is far enough and that the hottest point is no longer in the heating zone of the beam determined by the simulation.

$$\text{Speed} = \frac{(\text{Tmax}_{t\text{-}dt}\text{-Ttarget})^2}{0.072}\ \text{InitialSpeed} \quad (1)$$

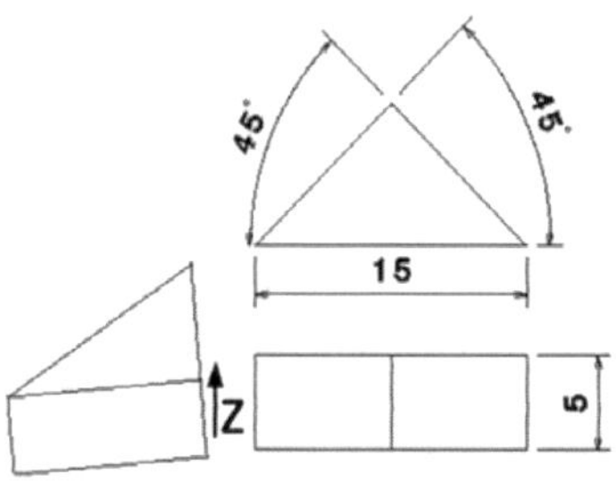

Figure 4. The case study part

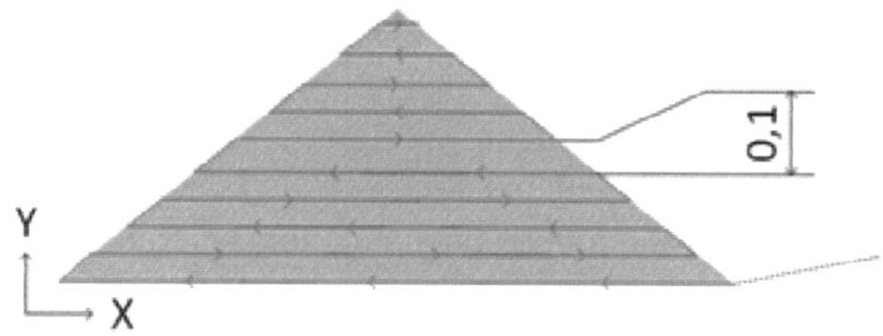

Figure 5. Melting trajectory for the prismatic part

The map of maximum temperatures reached resulting from the optimized melting strategy is shown in figure 6. This map can be compared with the maximum temperature reached with a constant speed equal to the initial speed and represented figure 7.

IOP Conf. Series: Journal of Physics: Conf. Series **885** (2017) 012016 doi:10.1088/1742-6596/885/1/012016

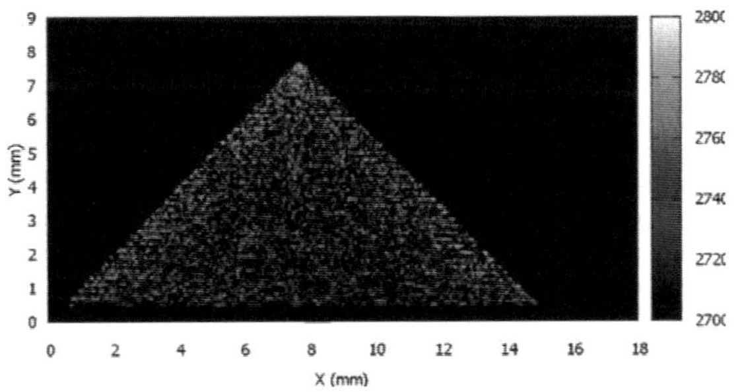

Figure 6. Maximum temperature reached with an optimised strategy

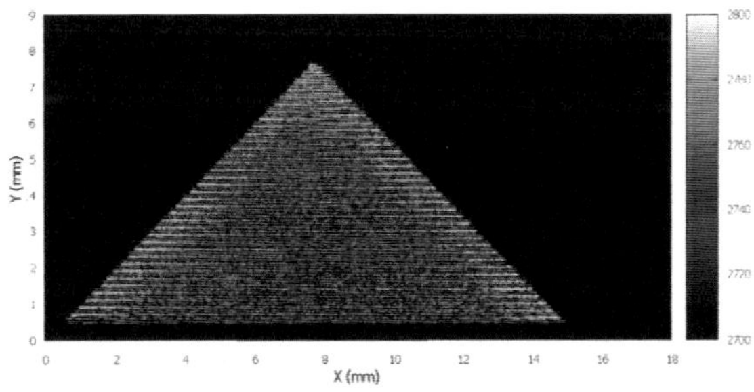

Figure 7. Maximum temperature reached with a constant speed

In order to have a finer analysis of the temperature distribution with or without optimization, the distribution of the voxel percentage as a function of the maximum temperature reached is plotted figure 8 for the constant speed strategy and figure 9 for the optimized strategy. Without optimization, the average temperature reached on the melted zone is 2596 °C whereas it is 2600 °C with optimization. This difference is not significant. On the other hand, the temperature distribution is substantially different. Without optimization (see figure 8), only 20.4% of the voxels in the melted zone reached a correct maximum temperature (between 2700 and 2760 ° C.). With the optimization (see figure 9), 57.0% of the voxels in the melted zone reached a correct maximum temperature. Even if this percentage can be further improved, the number of voxels likely to generate porosity or a defect in shape is thus greatly reduced by the use of a very simple optimization function.

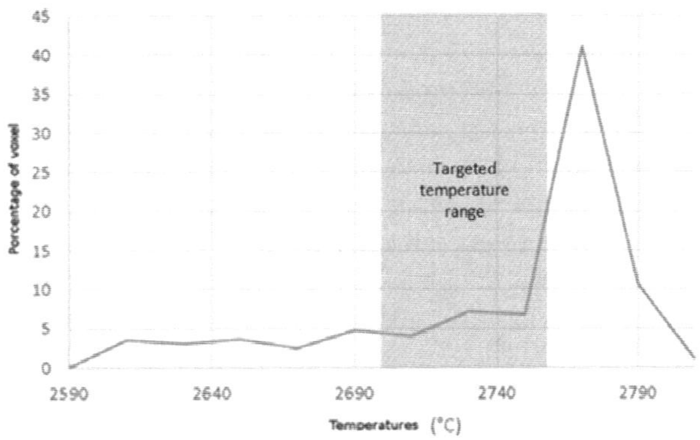

Figure 8. Maximum temperature repartition for constant beam speed

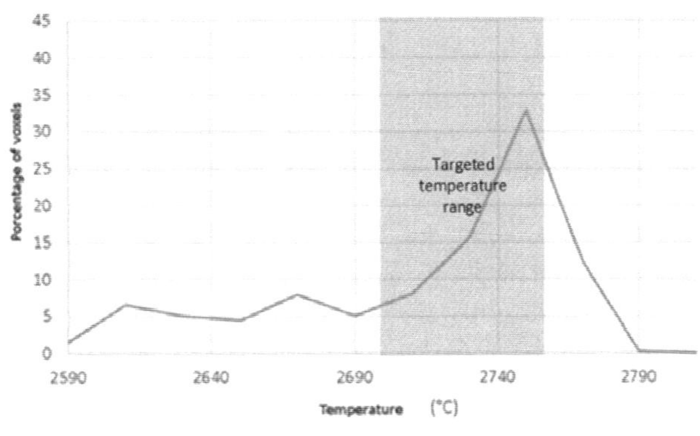

Figure 9. Maximum temperature repartition for optimized beam speed

Figure 10 shows that the speed of the beam increases when the beam passes back into an area already heated. This happen after a U-turn at the beginning of a new line. In this case the speed is increased from 1 to 1.2 m/s to avoid over heating of zone just exposed to the beam at the end of the previous scanned line. The speed is also increased when the lines are shorter near the top of the triangle. For example the line at y = 7.3 mm has a length of 0.6 mm. For this line, the average beam speed is 1,154 m/s as shown figure 11. For the line near the base of the triangle at y = 0.7 mmm and of length 14 mm, the average speed is 1.037 mm/s as shown figure 10. Close to the top of the triangle, the melted lengths are decreasing and the beam therefore returns more and more rapidly to areas already heated. Its speed must therefore be then increased to avoid overheating. Using simulation based melting strategy prevents overheating of the material while ensuring a minimum heating in order to avoid porosities.

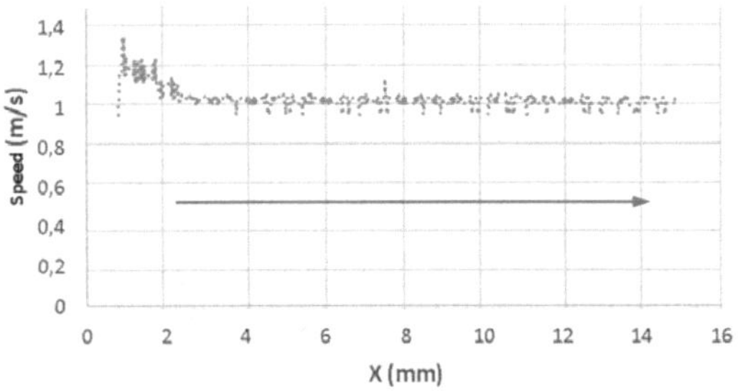

Figure 10. Speed along a long line

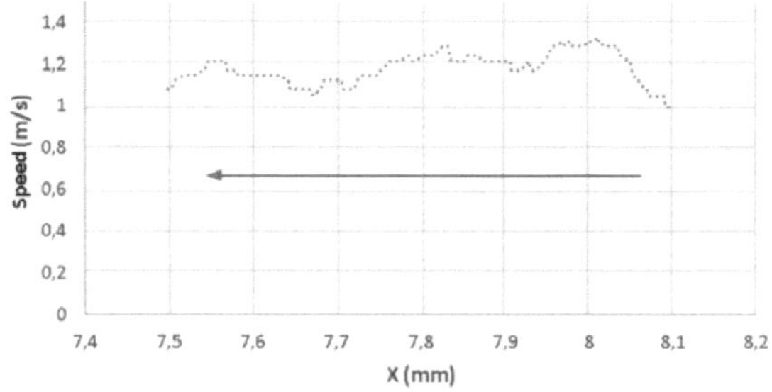

Figure 11. Speed along a short line

Conclusion

Defining optimal melting strategies is a key factor of part quality in additive manufacturing. This supposes first to be able to determine the key physical parameters. Then the relation between the melting strategies and theses key parameters have to be established. These two conditions being met, it becomes possible to predict the optimum melting conditions.

Chapter 3 of this paper highlighted the link between the temperatures reached by the material and its quality. Thus, recommendations on the temperatures to be reached in order to avoid defects in shape but also the porosities was stated and can be considered of the key physical parameter of part quality.

In chapter 4, a simulation-based optimization module was proposed to generate melting strategies that met the stated recommendations. In this paper a very simple optimization algorithm was used. Nevertheless, improving this algorithm, including more complex correction strategies, would allow to better meet the thermal criteria defined and further improve the parts quality. Moreover, in the long run, the optimization function should be able to optimize all the parameters of the fusion strategy and not only the speed of advance as is the case here. This also includes the trajectory followed by the electron beam.

References

[1] Shen N, Chou K. Simulations of Thermo-Mechanical Characteristics in Electron Beam Additive Manufacturing, American Society of Mechanical Engineers; 2012, p. 67–74. doi:10.1115/IMECE2012-88476.

[2] Shen N, Chou YK. Numerical thermal analysis in electron beam additive manufacturing with preheating effects. Proc. 23rd Solid Free. Fabr. Symp. Austin TX, 2012, p. 774–784.

[3] Vastola G, Zhang G, Pei QX, Zhang Y-W. Controlling of residual stress in additive manufacturing of Ti6Al4V by finite element modeling. Addit Manuf n.d. doi:10.1016/j.addma.2016.05.010.

[4] Kundakcioglu E, Lazoglu I, Rawal S. Transient thermal modeling of laser-based additive manufacturing for 3D freeform structures. Int J Adv Manuf Technol 2015:1–9.

doi:10.1007/s00170-015-7932-2.
[5] Pal D, Patil N, Zeng K, Stucker B. An Integrated Approach to Additive Manufacturing Simulations Using Physics Based, Coupled Multiscale Process Modeling. J Manuf Sci Eng 2014;136:061022–061022. doi:10.1115/1.4028580.
[6] Denlinger ER, Irwin J, Michaleris P. Thermomechanical Modeling of Additive Manufacturing Large Parts. J Manuf Sci Eng 2014;136:061007–061007. doi:10.1115/1.4028669.
[7] Kolossov S, Boillat E, Glardon R, Fischer P, Locher M. 3D FE simulation for temperature evolution in the selective laser sintering process. Int J Mach Tools Manuf 2004;44:117–23. doi:10.1016/j.ijmachtools.2003.10.019.
[8] Cheng B, Chou K. Geometric consideration of support structures in part overhang fabrications by electron beam additive manufacturing. Comput-Aided Des 2015;69:102–11. doi:10.1016/j.cad.2015.06.007.
[9] Béraud N, Vignat F, Villeneuve F, Dendievel R. Improving Dimensional Accuracy in EBM using Beam Characterization and Trajectory Optimization. Addit Manuf n.d. doi:10.1016/j.addma.2016.12.002.
[10] Körner C, Bauereiß A, Attar E. Fundamental consolidation mechanisms during selective beam melting of powders. Model Simul Mater Sci Eng 2013;21:085011. doi:10.1088/0965-0393/21/8/085011.
[11] Ammer R, Markl M, Ljungblad U, Körner C, Rüde U. Simulating fast electron beam melting with a parallel thermal free surface lattice Boltzmann method. Comput Math Appl 2014;67:318–30. doi:10.1016/j.camwa.2013.10.001.
[12] Markl M, Ammer R, Rüde U, Körner C. Numerical investigations on hatching process strategies for powder-bed-based additive manufacturing using an electron beam. Int J Adv Manuf Technol 2014;78:239–47. doi:10.1007/s00170-014-6594-9.
[13] Beraud N. Fabrication assistée par ordinateur pour le procédé EBM. Université Grenoble Alpes, 2016.
[14] Nguyen A-T, Reiter S, Rigo P. A review on simulation-based optimization methods applied to building performance analysis. Appl Energy 2014;113:1043–58. doi:10.1016/j.apenergy.2013.08.061.

IOP Conf. Series: Journal of Physics: Conf. Series **885** (2017) 012017 doi:10.1088/1742-6596/885/1/012017

Drying effects of vacuum far-infrared on aquilaria malaccensis leaves

Habsah Alwi, Sherif Abdulbari Ali, Ku Halim Ku Hamid, Mohamad Zaid Shamsudin, Nurhaslina Che Radzi, Nurul Asyikin Md Zaki & Miradatul Najwa Muhd Rodhi

Faculty of Chemical Engineering, Universiti Teknologi Mara, Shah Alam, 40450, Malaysia

habsahalwi@salam.uitm.edu.my

Abstract. Vacuum far-infrared radiation is an advanced drying technique that has the unique characteristics of energy transfer mechanism. In vacuum drying, the moisture evaporates at lower temperature while in infrared drying, the energy is absorbed directly which save a considerable amount of energy. The combination of the two would results in the time savings and enhancement of product quality. The effects of vacuum far-infrared drying on colour changes, elements and moisture content removal of *Aquilaria Malaccensis'* leaves were investigated. The parameter of the experiments were set to be temperature of 40, 50 and 60 °C, pressure of 0.6 and 0.8 bar, and a residence time of 120 minutes. The drying process reaches it optimum condition at temperature of 50°C and pressure of 0.6 bar, where the colour changes of the leaves, moisture content removal and elemental analyzer showed the best result of all six runs. The optimum drying conditions are recommended for further researches of the leaves.

1. Introduction

Aquilaria Malaccensis is a species of plant in the Thymelaeaceae family found primarily in Bangladesh, India, Indonesia, Laos, Malaysia, Myanmar, the Philippines, Singapore, and Thailand [1]. This species is of particular economic interest as it is the principal source of gaharu, one of the most highly valuable forest products currently traded internationally. The leaves of this species is believed to have a healing effects towards mild headaches and fever in some of the old and rural cultures that can be found from other natural herbs [2]. The many practices of agarwood has made the sustainable production of its kind become a major concern to prevent extinction [3]. The management in plantations of *Aquilaria* is recently established and it is foreseen that such plantings will deliver an alternative source of agarwood and lessen the harvesting pressures on its wild stands.

As of late, far-infrared (FIR) vacuum drying has been utilized as an advanced drying system because of various focal points. The uniqueness of using a vacuum far-infrared is that it lowers the pressure of the system and relatively reducing the operating temperature which in turns help in saving energy [4, 5]. Contrasted with other drying routines, FIR vacuum drying has a short drying time, simple establishment, high vitality proficiency, quick reaction of warming and drying and low capital expense [6]. The effectiveness of far-infrared (FIR) as a drying techniques onto the Aquilaria Subintegra leaves has discovered the potential FIR in preserving the antioxidant activity from the dried leaves extract [7].

Aquilaria Malaccensis has been known as one type of the popular and rare agarwood in the world. There are various types of active ingredients in almost every parts of the plant, including the leaves. The active ingredients extracted from the plant are used widely around the world. With its endangered population and long time to replanting, it is crucial to enhance its usage and preserve the active ingredients [8]. The leaves, often regards as waste or useless, are useful when treated and preserved accordingly. The aim of this study is to investigate the effects of vacuum far-infrared drying on *Aquilaria Malaccensis'* leaves, specifically on the colour changes, moisture content removal, and the elemental analysis of the leaves under VFIR conditions.

2. Methodology

2.1. Materials
Fresh leaves of *Aquilaria Malaccensis* were collected from a farm located in Jalan Kebun, Shah Alam prior to start this experiment. Particularly uniform size of *Aquilaria Malaccensis'* leaves with 3-5 cm in width and 6-10 cm in length that are matured and free from diseases were chosen to be used throughout the experiment. There were two batches of leaves collected which were in November and December of 2015, both came from the same farm and tree.

2.2. Collection of aquilaria malaccensis' leaves
The collected *Aquilaria Malaccensis'* leaves were cleaned by rinsing with running tap water, wiped dried with cleaned tissues to remove the impurities before let dried at room temperature in a laboratory. The cleaned leaves were then stored in a refrigerator at a temperature of 22°C and were ready for experimental procedures.

2.3. Experimental procedures
Six experimental run were selected, each with different parameters, to investigate the effect of VFIR drying on the leaves. The pilot scaled VFIR dryer used was fabricated by Nucleus Scientific & Instruments. The chamber inside the dryer was fitted with 3 FIR heat sources on three levels of the chamber. Temperature of the chamber is controlled using a system controller while the pressure is controlled using a vacuum pump. A set of temperature of 40, 50 and 60 °C while pressure of 0.6 and 0.8 bar were selected as the parameters in this experiment.
A handful amount of cleaned *Aquilaria Malaccensis'* leaves were taken for each run, placed on the tray inside the VFIR, five pieces almost identical leaves were chosen randomly to be placed on the weighing balance and be used as sample for the colour changes analysis as well as elemental analysis before and after each run. Each run was performed for 120 minutes so that the leaves were well-dried in a fixed residence time.

2.4. Colour measurement
For colour changes analysis, the leaves' colour was measured using a Konica Minolta Chromameter (B8208418). Each of the five selected leaves were measured of their colour using the chromameter before and after each run were performed, with all five leaves were analyzed at the same spot for both before and after each run, in order to get the most accurate results. Hunter scale values of L, a, and b were used where L=0 for black, L=100 for white, a<0 for green, a> for red, b<0 for blue, and b>0 for yellow. Chroma (C), which indicate the colour intensity, was calculated as below:

$$C= a^2 + b^2 \tag{1}$$

The hue angle (h) from 0 or 360° indicates a red hue and 90, 180, and 270° indicate yellow, green, and blue hues, respectively. The h value was calculated as below :

$$h = \tan^{-1} b \ a \tag{2}$$

2.5. Moisture content removal

For the moisture content removal of the leaves after drying, data were collected from the weighing balance installed within the VFIR dryer itself and the weight of the five leaves placed upon the balance were taken before the dryer is turned on and after 120 minutes of drying. The difference in weight would indicate the total moisture loss of the five leaves from the drying process.

2.6. Elemental analysis

The elemental analysis of the leaves before and after the drying process were conducted using a Thermo Fisher Scientific Flash EA 1112 elemental analyser. The leaves that need to be analysed were ground until the size of almost powder because the elemental analyser can only analysed sample that is in very little size.

Twelve samples were analysed prior to six experimental run with two samples each. The results were given in the percentage concentration of nitrogen, carbon, and hydrogen atoms in the sample, where comparison between the composition in before and after drying were made.

3. Results and discussion

3.1. Effect of VFIR drying on the colour changes of the leaves

The effect of VFIR drying on *Aquilaria Malaccensis'* leaves were investigate in terms of the colour changes of the leaves after it has been dried using the chromameter. Fresh *Aquilaria Malaccensis's* leaves have a green and slightly yellow colour before it turned into dark yellow after it has been dried. Colour is an important physicochemical property. The impact of VFIR drying conditions on colour parameters of L, C and h of dried *Aquilaria Malaccensis'* leaves were examined and results are shown in Table 1. *Aquilaria Malaccensis'* leaves dried at 60°C and 0.6bar (run 6) had the highest L value which is 42.93, while *Aquilaria Malaccensis'* leaves dried at 60°C and 0.8bar (run 5) had the highest C value, 19.69. The h value indicated that all dried *Aquilaria Malaccensis'* leaves were yellow and slightly green. The results obtained from this experiment follows the previous study in the increment of values, except the hue angle. This may be due to the fact that the type of leaves used is different.

Table 1. Effect of VFIR drying on the colour of Aquilaria Malaccensis' leaves

Run	Variable		Responses								
	T (°C)	P (bar)	Lightness			Chroma			Hue Angle		
			Before	After	% Difference	Before	After	% Difference	Before	After	% Difference
1	40	0.8	35.24	39.34	11.62	16.33	17.36	6.29	43.68	66.69	52.67
2	40	0.6	34.94	39.94	14.30	16.64	17.67	6.16	48.02	66.04	37.54
3	50	0.8	36.32	41.84	15.22	18.79	19.10	1.64	47.53	71.23	49.86
4	50	0.6	35.21	40.63	15.41	15.24	18.15	19.11	44.61	67.53	51.36
5	60	0.8	35.28	41.86	18.64	17.37	19.69	13.34	49.26	73.48	49.17
6	60	0.6	36.05	42.93	19.10	17.86	18.90	5.79	45.93	72.32	57.44

From the table 1, it can be seen that run 4 (50°C and 0.6 bar) shows a high percentage differences of the total three categories namely lightness (L), chroma (C) and hue angle (h). On the contrary, run 2 (40°C and 0.6 bar) shows the lowest total percentage of differences. This is due to the lower temperature and greater vacuum has been achieved.

In previous study using Miang leaves [9], the optimum temperature was found to be at 65°C and 0.53 bar for the drying of Miang leaves with the lightness and chroma of the leaves increased after the drying process, but the hue angle only had a minor increment. The results obtained from the experiment follows the previous study in the increment of values, except the hue angle. This may be due to the fact that the type of leaves used is different.

3.2. Effect of VFIR drying on the moisture content removal of the leaves

The drying process of the leaves using VFIR dryer has caused some weight loss of the leaves after the drying and this is due to the moisture loss inside the leaves itself. In any drying process, moisture content inside the sample will be loss and it is the sole purpose of drying process to be conducted on any sample. Drying of leaves were conducted to ensure moisture loss and preserved the ingredients inside the leaves. For this experiment, the moisture content removal of the five leaves placed on the weighing balance were determined and tabulated in Table 2.

The highest percentage of moisture content removal among the six run is run 5 using 60°C and 0.8 bar with 8.55% loss. The trend of the moisture content removal keeps on increasing until run 5, which then shows a little decrement of only 0.09% in run 6. This may be of the factor that at run 5 (60°C and 0.8 bar), it is the optimum temperature and pressure for drying Aquilaria Malaccensis' leaves which explains why at run 6 (60°C and 0.6 bar), a decrease in the pressure did not increase the percentage of moisture content removal of the leaves.

In a study of carrot's drying using VFIR dryer [3], an increase in the temperature from 70 to 80°C does not showed a promising results and thus conclude that for each type of sample, the increase of temperature does not always mean the moisture content removal will be better.

Table 2. Effect of VFIR drying on the moisture content removal of Aquilaria Malaccensis' leaves

Run	Variable		Responses		
	T	P	Weight of leaves		
	(°C)	(bar)	Before	After	% Difference
1	40	0.8	29.84	28.63	4.05
2	40	0.6	30.19	28.84	4.47
3	50	0.8	30.05	27.82	7.42
4	50	0.6	30.64	28.04	8.49
5	60	0.8	29.34	26.83	8.55
6	60	0.6	29.78	27.26	8.46

3.3. Effect of VFIR drying on the elemental composition of the leaves

The purpose of analyzing the elemental composition of the fresh and dried *Aquilaria Malaccensis'* leaves was to compare the composition of nitrogen, carbon and hydrogen inside the leaves before and after it was dried using the VFIR dryer. The results of the analysis are tabulated in Table 3 below.

The process of drying requires heat and along the way, oxygen will be used in the process, making oxygen atoms and molecules less available inside the sample [10]. This explains why the composition of nitrogen, carbon and hydrogen increased after the drying process has stopped, because of the depletion of oxygen molecules. There may be other atoms that are presents in the sample but this experiment only covers the composition of nitrogen, carbon and hydrogen.

Table 3. Effect of VFIR drying on the elemental composition of Aquilaria Malaccensis' leaves

Run	Variable		Responses								
			Nitrogen			Carbon			Hydrogen		
	T (°C)	P (bar)	Before	After	% Difference	Before	After	% Difference	Before	After	% Difference
1	40	0.8	2.04	2.11	3.43	36.39	41.92	15.20	6.18	6.32	2.27
2	40	0.6	1.98	2.06	4.04	37.44	41.77	11.57	5.99	6.28	4.84
3	50	0.8	1.86	1.96	5.38	36.68	42.53	15.95	6.52	6.65	1.99
4	50	0.6	2.01	2.19	8.96	32.48	42.05	29.46	6.20	6.40	3.23
5	60	0.8	1.93	2.06	6.74	34.62	42.15	21.75	6.27	6.58	4.94
6	60	0.6	1.87	2.04	9.09	35.37	43.21	22.17	6.19	6.41	3.55

As shown in the Table 3, run 5 (50°C and 0.6 bar) obtained the highest percentage differences of all three elements among the six while run 2 continues to obtain the least percentage differences. This shows that the drying process is most effectively conducted at temperature 50°C and pressure of 0.6 bar.

4. Conclusion
The effects of VFIR drying on the *Aquilaria Malaccensis's* leaves were investigated on three different aspects namely; colour changes, moisture content removal and elemental analysis of the leaves before and after drying. All the data and observations throughout the experiment has pointed out that the best drying conditions for the *Aquilaria Malaccensis's* leaves are at temperature of 50°C and pressure of 0.6 bar. This is to ensure that the leaves are well-dried and the active ingredients inside it could be preserve for future uses and researches. At this condition, the moisture content removal although not at the highest among all 6 runs came in closed second with 8.49% difference, just 0.06% behind the highest moisture content removal, while the colour changes and elemental composition both showed the best results.

Acknowledgements
The author would like to thanks Ministry of Education (MOE), Malaysia for providing the financial support through Fundamental Research Grant Scheme (FRGS), 600-RMI/FRGS 5/3 (95/2013) for this research and Faculty of Chemical Engineering, Universiti Teknologi MARA for the research facilities.

References
[1] Benedict A. C., (2009). Extraction of the Essential Oil of *Aquilaria Malaccensis* (Gaharu) Using Hydro-Distillation and Solvent Extraction Methods. Faculty of Chemical & Natyral Resources Engineering, Universiti Malaysia Pahang
[2] Afiffudden, S. K. N, Alwi, H & Ku Hamid, K. H., (2015) Potential Antipyretic Agents From Plants : An Overview., Advanced Materials Research, Volume 1113, pp 382-387
[3] Botanic Gardens Conservation International (2008). https://www.bgci.org/resources/article/0576/, Retrieved *13th April 2016.*
[4] Mongpraneet S., Abe T., & Tsurusaki T., (2002), Accelerated drying of welsh onion by far infrared radiation under vacuum conditions., *Journal of Food Endineering, 55, pp 147-156*
[5] Swasdisevi T., Dehasvatin S., Sa-Adchom P., & Soponronnarit S.,(2009), Mathematical Modeling of Combined Far-Infrared and Vacuum Drying Banana Slice. *Journal of Food Engineering, 92, pp 100-106*
[6] C. Nimmol (2010). Vacuum Far-infrared Drying of Foods and Agricultural Materials. Journal of KMUTNB, Vol 20, No.1
[7] H Alwi, M Zaki, N Asyikin, N Che Radzi, M Rodhi, M Najwa, MA Ami, KHK Hamid, (2017) Heat Effects from Far-Infrared Source towards Antioxidant Activity in Aquilaria subintegra Leaves, Materials Science Forum 890, 146-149
[8] Afiffudden, S. K. N, Alwi, H & Ku Hamid, K. H., (2015) Determination of 4'-Hydroxyacetanilide in Leaves Extract of Aquilaria malaccencis by High Pressure Liquid Chromatograph, Procedia - Social and Behavioral Sciences, Vol 195, pp 2726-2733
[9] Hirun S.,Choi J., Choi J., Ayarungsaritkul J., Ayarungsaritkul J., Pawsaut C., Pawsaut C., Sutthiwanjampa C., Sutthiwanjampa C., Vuong Q., Chockchaisawasdee S., Heo Y., & Scarlett C., (2015). Optimization of Far-infrared Vacuum Drying Conditions for Miang Leaves (Camellia sinensis var. assamica) Using Response Surface Methodology. Food Science Biotechnology Journal, 24(2), pp 461-469.
[10] Alfani A., Maisto G., Lovieno P., Rutigliano F. A., & Bartoli G. (1996), Leaf contamination by atmospheric pollutants as assessed by elemental analysis of leaf tissue, leaf surface deposit and soil. Journal of Plant Physiology, 148, pp 243-248

A numerical solution for the axisymmetric flow of a plastic tube on a rigid fibre

Thanh Nguyen[1], Kien Nguyen[2], Sergei Alexandrov[3]

[1] Institute of Mechanics – VAST, Ha Noi, Viet Nam
E-mail: manhthanh2012209@gmail.com
[2] University of Communications and Transport, Ha Noi, Viet Nam
E-mail: ntkien@utc.edu.vn
[3] Institute for Problems in Mechanics - Russian Academy of Sciences
Moscow, Russia. E-mail: sergei_alexandrov@yahoo.com

manhthanh2012209@gmail.com

Abstract. In the case of rigid perfectly plastic material the strain rate intensity factor appears in the vicinity of maximum friction surfaces. This factor controls the magnitude of the equivalent strain rate in a narrow region near the surface. On the other hand, the equivalent strain rate controls the evolution of material surfaces. Therefore, the existence of the strain rate intensity factor in theoretical solutions are in qualitative agreement with experimental observations that demonstrate that a narrow layer with drastically modified microstructure is often generated in the vicinity of frictional interfaces in machining and deformation processes. In order to use the strain rate intensity factor for describing such material behaviour, it is necessary to develop an efficient numerical method for calculating the strain rate intensity factor. Since this factor is involved in a singular series expansion, it is evident that commercial finite element packages are not capable of calculating the strain rate intensity factor. In the case of Tresca's solids the theory of characteristics can be used for the development of the method in question. This paper presents a first step in this direction

Keywords: strain rate intensity factor, rigid perfectly plastic material, axisymmetric flow, total plasticity

1. Introduction

Numerous experimental observations demonstrate that a narrow layer with drastically modified microstructure is generated near frictional interfaces in machining and deformation processes. This layer is usually called white layer in papers devoted to machining processes and fine grain layer in papers devoted to metal forming processes. A complete review of results on white layer/fine grain layer generation published before 1987 has been presented in [1]. According to this review article there are three main contributory mechanisms responsible for white layer generation. One of these mechanisms is the mechanism of intensive plastic deformation. This mechanism can be described by means of the strain rate intensity factor introduced in [2]. This factor is the coefficient of the leading singular term in a series expansion of the equivalent strain rate near maximum friction surfaces. This expansion shows that the equivalent strain rate is infinite at the friction surface. Therefore, the strain rate intensity factor controls the magnitude of the equivalent strain rate in its vicinity. On the other hand, the equivalent strain rate controls the evolution of material properties. In the case of rigid perfectly plastic materials a necessary condition for the existence of the strain rate intensity factor is the friction boundary condition that postulates that the friction stress at sliding is equal to the shear yield stress (maximum friction law). This boundary condition is often adopted at the tool - chip interface (at least, over a portion of this interface) in machining processes [3 - 10]. This zone is usually called the sticking zone. The existence of such zones has been reported in deformation processes as well [11 - 14]. However, it is worthy of note that the actual seizure (the velocity vector is continuous across the interface) may or may not occur in theoretical solutions. It depends on the constitutive equations chosen. Also, the interpretation of experimental results on this issue is controversial [15]. In the present paper, it is assumed that sliding occurs at maximum friction surfaces. In this case models for describing the generation of fine grain layers near frictional interfaces based on the strain rate intensity factor have been proposed in [16, 17]. In order to apply these models to industrial problems, it is necessary to carry out a special experimental program and to develop an accurate method for calculating the stain rate intensity factor. Since this factor is involved in a singular asymptotic expansion of the equivalent strain rate, it is evident that commercial finite elements packages are not capable of calculating the strain rate intensity factor. Probably, the generalized finite element method [18] can be used for this purpose. However, no specific code is currently available. To the best of authors' knowledge, the only available numerical approach for calculating the strain rate intensity factor is based on the method of characteristics [19]. This approach is valid for plane strain deformation of rigid perfectly plastic material. On the other hand, most of experimental results are available for axisymmetric deformation, for example [20, 21]. Using analytic and semi-analytic solutions the strain rate intensity factor in axisymmetric flows has been determined in [22, 23]. However, comparison of the accurate numerical plane strain solution derived in [19] and the corresponding approximate analytic solution demonstrates that the accuracy of the analytic solution is not sufficient for practical applications. Therefore, an efficient numerical method for calculating the strain rate intensity factor in axisymmetric flow is needed. In the case of Tresca's solids (i.e. solids satisfying Tresca's yield criterion and its associated flow rule) such a method can be based on the theory of characteristics and can be considered as an extension of the method developed in [19]. In particular, the general theory of characteristics for this case is presented in [24]. In the present paper, as a first step to the development of a numerical method for calculating the strain rate intensity factor in axisymmetric flows, this theory is used in the conjunction with the finite difference method to build up the field of characteristics and, as a result, the field of stress for the problem formulated in [23] but without using simplified assumptions accepted in this paper. Then, comparison of the numerical and analytic solutions is made.

2. Boundary value problem and constitutive equations

Consider the compression of rigid perfectly plastic tube on a rigid fibre of radius a (Fig. 1). The outer radius of the tube is b and its inner radius is a. The length of both the fibre and tube is $2L$. Pressure applied at the outer surface of the tube causes axisymmetric plastic deformation in it. Cylindrical coordinate system (r, θ, z) is taken such that the z-axis coincides with the axis of symmetry of the flow and the plane $z = 0$ is the plane of symmetry. Then, it is sufficient to consider the region $0 \le z \le L$. The non-zero stress components are denoted by σ_r, σ_θ, σ_z, and τ_{rz}. Velocity components in the radial and axial directions are denoted by v_r and v_z respectively. The boundary conditions are

$$\tau_{rz} = 0 \quad \text{at} \quad r = b, \tag{1a}$$

$$\tau_{rz} = k \quad \text{at} \quad r = a, \tag{1b}$$

$$\sigma_z = 0 \text{ and } \tau_{rz} = 0 \quad \text{at} \quad z = L \tag{1c}$$

$$v_r = -U \quad \text{at} \quad r = b, \tag{2a}$$

$$v_r = 0 \quad \text{at} \quad r = a. \tag{2b}$$

IOP Conf. Series: Journal of Physics: Conf. Series **885** (2017) 012018 doi:10.1088/1742-6596/885/1/012018

here k is shear yield stress. On the plane $z = 0$, by symmetry

$$v_z = 0 \quad \text{and} \quad \tau_{rz} = 0 \tag{3}$$

In the present problem, the case of interest is that in which $v_r \leq 0$. Then, under conditions of axial symmetry, it is assumed that the material of the tube satisfies Tresca's yield criterion in the regime of total plasticity (see, for exemple [24, 25]), so that

$$(\sigma_r - \sigma_z)^2 + 4\tau_{rz}^2 = 4k^2 \quad ; \quad \sigma_\theta = \frac{1}{2}(\sigma_r + \sigma_z) - k \tag{4}$$

The stress components satisfy the equilibrium equations

$$\frac{\partial \sigma_r}{\partial r} + \frac{\partial \tau_{rz}}{\partial z} + \frac{\sigma_r - \sigma_\theta}{r} = 0 \tag{5a}$$

$$\frac{\partial \tau_{rz}}{\partial r} + \frac{\partial \sigma_z}{\partial z} + \frac{\tau_{rz}}{r} = 0 \tag{5b}$$

and, there are four equations for the determination of the four unknown stress components, i.e. the problem is statically determinate.

The velocity components are determined by the incompressibility and isotropy conditions

$$\frac{\partial v_r}{\partial r} + \frac{v_r}{r} + \frac{\partial v_z}{z} = 0 \tag{6}$$

$$\left(\frac{\partial v_r}{\partial z} + \frac{\partial v_z}{\partial r}\right) / \left(\frac{\partial v_r}{\partial r} - \frac{\partial v_z}{\partial z}\right) = 2\tau_{rz} / (\sigma_r - \sigma_z) \tag{7}$$

It is known that these systems (i.e. (5a)-(5b) and (6)-(7)) are hyperbolic and the characteristics for the stresses and the velocities coincide, furthermore, they coincide with the slip-lines. Therefore, there are two distinct characteristic directions at a point, denoted by α and β respectively. In such a case the method of characteristics can be used to solve the problem under consideration. Substituting

$$\sigma_r , \sigma_z = \sigma \pm k \sin 2\varphi , \quad \tau_{rz} = k \cos 2\varphi \tag{8}$$

here $\sigma = (\sigma_r + \sigma_z)/2$ and φ is the anti-clockwise angular rotation of the α-line from the z-axis. Then, equations for α-lines and β-lines are

$$dr/dz = \tan\varphi \quad \text{and} \quad dr/dz = -\cot\varphi \tag{9}$$

Now the α-line and β-line are regarded as a pair of right-handed curvilinear axes of reference and denoted by s_α and

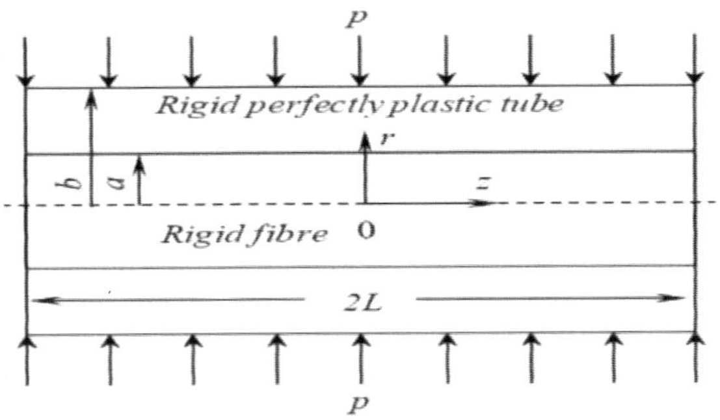

Figure 1. Configuration and coordinate system.

s_β respectively. Then, following [24] in transforming from the coordinate system (z, r) to the characteristic coordinates (s_α , s_β), with v_α and v_β being the components of the velocity vector along the characteristic coordinates, the equations (5a), (5b), (6) and (7) take the form

$$d\sigma - 2k d\varphi = \frac{k}{r}(-dr - dz) \text{ along an } \alpha\text{-line} \tag{10a}$$

$$d\sigma + 2k d\varphi = \frac{k}{r}(-dr + dz) \text{ along a } \beta\text{-line} \tag{10b}$$

IOP Conf. Series: Journal of Physics: Conf. Series **885** (2017) 012018 doi:10.1088/1742-6596/885/1/012018

$$dv_\alpha - v_\beta d\varphi = -\frac{1}{2r}(v_\alpha dz - v_\beta dr) \text{ along an } \alpha\text{-line} \tag{11a}$$

$$dv_\beta + v_\alpha d\varphi = -\frac{1}{2r}(v_\alpha dr + v_\beta dz) \text{ along a } \beta\text{-line} \tag{11b}$$

and the boundary conditions (1a-b-c) and (2a-b) become

$$\varphi = 7\pi/4 \quad \text{at} \quad r = b, \tag{12a}$$

$$\varphi = 2\pi \quad \text{at} \quad r = a, \tag{12b}$$

$$\sigma = -k \quad \text{and} \quad \varphi = 7\pi/4 \quad \text{at} \quad z = L \tag{12c}$$

$$v_\beta = v_\alpha - U\sqrt{2} \quad \text{at} \quad r = b, \tag{13a}$$

$$v_\beta = 0 \quad \text{at} \quad r = a. \tag{13b}$$

3. Numerical solution

3.1. Numerical scheme for the problem of Sect. 2

Small displayed equations: Some examples: Several numerical schemes based on the method of characteristics have been proposed to determine the stress and velocity distributions in axisymmetric flow of rigid perfectly plastic solids [24]. In the present paper we adopt one of these scheme, which is analogue to the scheme presented in [19, 26] for the problem in plane strain flow, to solve the problem formulated in the previous section.

The general structure of the characteristics and Cartesian coordinates (z, r) are symbolically illustrated in Fig. 2. Starting from base-line AB, the stress distribution across a network of characteristics (slip-lines) can be uniquely defined by using the systems (9), (10a-b), and the boundary conditions (12a-b-c). The equations (9), (10a-b) (expressed in finite-difference form) are nonlinear and solved by suitable iterative procedures.

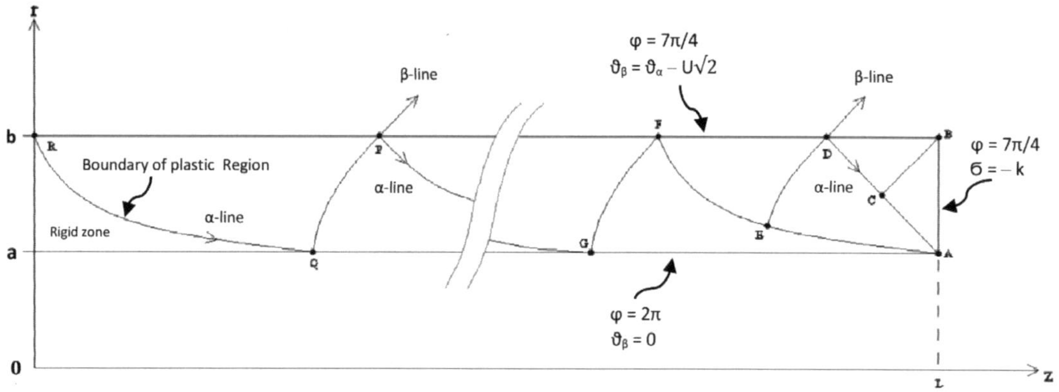

Figure 2. Configuration and coordinate system ($z \geq 0$, $r \geq 0$).

The stress solution and slip-line field in the region ABC are determined by the conditions (12c) at N equally spaced nodal points on the segment AB. The nodal points on β-line BC were taken as the intersections with the α-lines (through the nodal points on AB). The nodal points on α-line AC were taken in a similar way. Next, the field BCD is defined by the solutions on BC and the boundary condition (12a). The point A is a singularity through which pass all α-lines within an angle of 45° between AD and AG. Having the solutions on AD, the field ADE is defined by the numerical procedure proposed in [24, 27]. The construction of the field continues in similar way up to the slip-line through the point R located on the intersection of the plane $z = 0$ with the outer surface of the tube. This curve is the α-line and also the boundary between the rigid and plastic regions.

Having completed the slip-line field, the velocity solution is built up from the left in Fig. 2 and uniquely determined by the conditions for the velocities on BR, AQ and RQ. At this moment a suitable velocity boundary conditions should be set on the rigid plastic boundary RQ. The wedge shaped area included between RQ and its counterpart in the left-hand quadrant is rigid and do not move down as a whole body but lose material to the plastic region. Using (11a), (12a) and taking into account that the normal component of velocity vector is continuous across RQ yield

$$v_\beta = 0 \ , \quad v_\alpha = U.\sqrt{2b/r} \quad \text{on} \quad RQ \tag{14}$$

It is evident that the tangential component of velocity v_α is discontinuous across RQ.
Now that all the velocity boundary conditions have been identified, starting from rigid plastic boundary RQ, the velocity solution can be found using linear system of equations (11a-b) in finite-difference form. The construction of the velocity field is almost identical to the one used in the stress determination.

3.2. Numerical results and comparisons.
An analytic solution to the problem formulated in the section 2 has been given by Spencer A. J. M. in [25]. This solution is approximate because the boundary conditions at $z = 0$ and $z = L$ are ignored. However, all the equations and the boundary conditions at $r = a$ and $r = b$ are satisfied. Since Spencer's solution gives a good approximation to the solution for compression of rigid perfectly plastic tube on a rigid fibre, except near the ends and center of the tube, it is inferred that Spencer's solution is appropriate to verify numerical results obtained here if the length of the tube is large enough compared to its radius. Note that when calculating the stress and velocity fields by Spencer's formula, the following boundary conditions in the integral form have been used [23]

$$\int_a^b r\sigma_z|_{z=L}dr = 0, \ \int_a^b rv_z|_{z=0}dr = 0 \tag{15}$$

Without loss of generality, it is assumed that U = 1 and (b – a) = 1. The calculations were performed for tubes with a ratio L / (b – a) in the range of 5 – 25, and an inner radius a in the range of 1–20. Fig. 3 shows two slip-line field configurations corresponding to {L = 5.54, a = 1} and {L = 6.25, a = 3} respectively. In both cases, the number of nodes N on the segment AB are equal to 6. It is seen from Fig. 3 that both families of characteristics are straight in the triangle ABD in which the two relationships are satisfied : dr = – dz along an α-line and dr = dz along a β-line, therefore all the stress components are constant by (8) and (10a-b). Thus, the stress state is uniform in the region ABD.
Next, the radial stress and axial velocity distributions on AQ (along the friction surface r = a) for the both numerical and analytic solutions are compared in Fig. 4. With N = 11, the calculations were performed for the tube with L = 13.86, a = 1. In this case, as expected, the numerical and analytic solutions for σ_r and v_z are very closed to each other when the variable z lies in the distances greater than 2(b – a) from the ends and greater than 3(b – a) from the center of the tube.

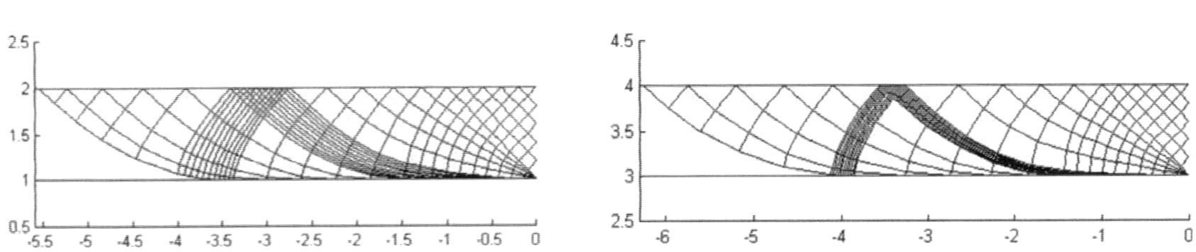

L = 5.53 , a = 1 L = 6.25 , a = 3

Figure 3. Slip-line fields calculated for tubes of diferent radius and length-diameter ratio (N = 6).

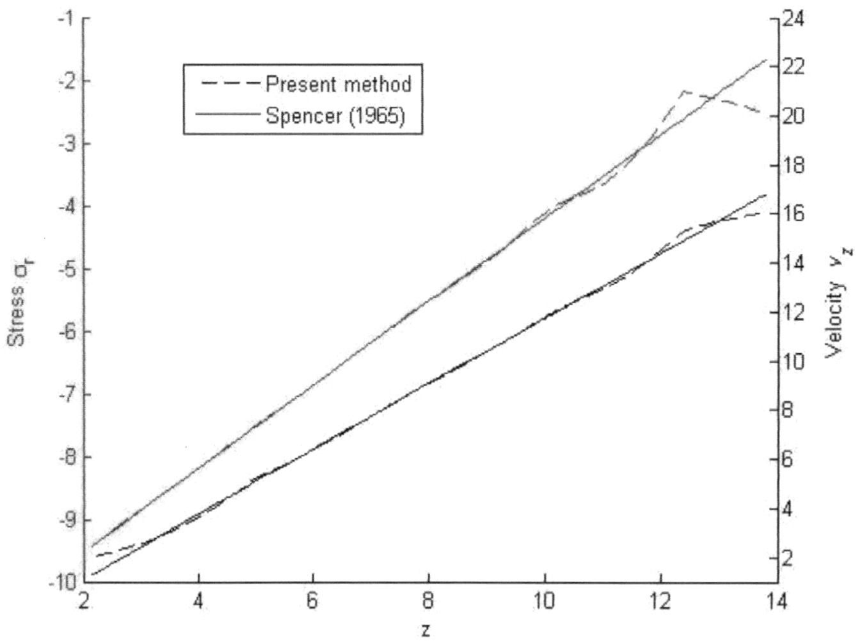

Figure 4. Comparison of numerical and analytical values of σ_r , v_z along the friction surface $r = a$ ($L = 13.86$, $N = 11$).

Considering that Spencer's solution is accurate, at least in places situated at a long enough distance from the center and the ends of the tube, it is reasonable to introduce the following definitions for the investigation of the distribution of relative error within the plastic region:

$\boldsymbol{\beta}$ - a set of all the nodal points on the β-line, which is selected for comparison.

$\boldsymbol{\beta}_{L|Za}$ - a set of all the nodal points on the β-line passing through point ($z = z_a$, $r = a$).

$\boldsymbol{\delta}(\boldsymbol{f}\,|\,\boldsymbol{\beta})$ - the maximum absolute value of the relative error of the value f in the set β

$$\boldsymbol{\delta}(\boldsymbol{f}|\boldsymbol{\beta}) = \max_{(z_i,r_i)\in\boldsymbol{\beta}} \left|\frac{\boldsymbol{f}^{(A)}(z_i,r_i) - \boldsymbol{f}^{(N)}(z_i,r_i)}{\boldsymbol{f}^{(A)}(z_i,r_i)}\right| \quad (16)$$

here $\boldsymbol{f}^{(A)}(z_i,r_i)$ and $\boldsymbol{f}^{(N)}(z_i,r_i)$ are the calculated values of Spencer's and numerical solutions respectively at (z_i,r_i).

Now there are three β-lines selected within the plastic region for each tube. The first passing through point (z_a , a) located close to the end of the tube, but $z_a < L - 2(b - a)$. The second passing through the point located near the middle of AQ. The third passing through point (z_a , a) located close to the center of the tube, but $z_a > z_Q$ (the z-coordinate of the point Q). The positions of these β-lines are relatively specific to the distribution of the calculated values. Having calculated the stress and velocity fields (for two cases $L = \boldsymbol{L1} = 13.86$, $L = \boldsymbol{L2} = 22.2$ with $N = 11$) according to Spencer's and numerical solutions, the maximum relative errors of all the calculated values on the above selected β-lines are found from (16) and presented in Tab. I. It is seen from this table that the relative errors are very small for the middle β-lines despite the number N is relatively small. Also, the error decreases as the length-diameter ratio increase over plastic zone except the ends.

Table1. Relative Errors of Numerical values.

Set β	Maximum Value of The Relative Error				
	$\delta(\tau_{rz}\|\beta)$	$\delta(\sigma_r\|\beta)$	$\delta(\sigma_z\|\beta)$	$\delta(v_r\|\beta)$	$\delta(v_z\|\beta)$
$\beta_{L1}\|_{11.82}$	3.9e-1	2.9e-3	5.6e-2	2.2e-1	5.2e-3
$\beta_{L2}\|_{20.2}$	7.0e-2	3.4e-2	1.3e-1	2.7e-1	5.0e-3
$\beta_{L1}\|_{7.7}$	2.7e-2	2.1e-3	1.4e-3	1.0e-2	1.9e-4
$\beta_{L2}\|_{11.21}$	1.9e-4	6.5e-4	8.4e-4	7.5e-4	4.2e-3
$\beta_{L1}\|_{2.83}$	3.6e-4	5.9e-4	7.8e-4	3.0e-1	2.5e-2
$\beta_{L2}\|_{3.07}$	5.5e-4	5.6e-4	6.4e-4	8.4e -2	0.9e-5

4. Conclusion

A numerical approach for determining the stress and velocity fields of rigid perfectly plastic material in axisymmetric flows of a plastic tube on a rigid fibre has been performed. The approach is based on the method of characteristics. The accuracy of the numerical results has been verified by comparison with an accurate analytic solution. If the length-diameter ratio of the tube is large enough, the magnitude of the stress and velocity components at the places situated in the distances by 2-3 times the tube thickness from its ends and center is very close to that found from the approximate solution in which the end effects are neglected.
Since the determination of the strain rate intensity factor is based on the derivation of the equivalent strain rate in the vicinity of an envelope of characteristics, the calculation of the velocity and characteristic field with high accuracy is of high importance, therefore the numerical approach presented in the present paper is the first and right step to the development of a numerical method for calculating the strain rate intensity factor in axisymmetric flows.

Acknowledgments
The research described has been supported by the grants VAST.HTQT.Nga.02/16-17 (Vietnam) and RFRB-16-51-540001 (Russia).

References

[1] Griffiths B.J. (1987) Mechanisms of White Layer Generation with Reference to Machining and Deformation Processes, Trans. ASME J. Tribol, 109, 525-530.
[2] Alexandrov S., Richmond O. (2001) Singular Plastic Flow Fields near Surfaces of Maximum Friction Stress, Int. J. Non-Linear Mech. 36, 1-11.
[3] Ng E.-G., Aspinwall D.K., Brazil D., Monaghan J. (1999) Modelling of Temperature and Forces When Orthogonally Machining Hardened Steel, Int. J. Mach. Tool Manu., 39, 885-903.
[4] Yiğit K., Tuğrul Ö. (2006) Predictive Analytical and Thermal Modeling of Orthogonal Cutting Process—Part I: Predictions of Tool Forces, Stresses, and Temperature Distributions, J. Manu. Sci. Eng., 128, 435-444.
[5] Ramesh A., Melkote Shreyes N. (2008) Modeling of White Layer Formation under Thermally Dominate Condition in Orthogonal Machining of Hardeded AISI 52100 Steel, Int. J. Mach. Tool. Manu., 48, 402-414.
[6] Lalwani D.I., Mehta N.K., Jain P.K. (2009) Extension of Oxley's Predictive Machining Theory for Johnson and Cook Flow Stress Model, J. Mater. Process. Tech., 209, 5305-5312.
[7] Akbar F., Mativenga Paul T., Sheikh M.A. (2010) An Experimental and Coupled Thermo-Mechanical Finite Element Study of Heat Partition Effects in Machining, Int. J. Adv. Manuf. Technol. 46, 491-507.
[8] Molinari A., Cheriguene R., Miguelez H. (2012) Contact Variables and Thermal Effects at the

Tool-chip Interface in Orthogonal Cutting, Int. J. Solids. Struct., 49, 3774-3796.

[9] Chen G., Li J., He Y., Ren C. (2014) A New Approach to The Determination of Plastic Flow Stress and Failure Initiation Strain for Aluminium Alloys Cutting Process, Comp. Mater. Sci., 95, 568-578.

[10] Agmell M., Ahadi A., Stahl J.-E. (2014) Identification of Plastic Constants from Orthogonal Cutting and Inverse Analysis, Mech. Mater., 77, 43-51.

[11] Kim T.-K., Ikeda K. (2000) Flow Behavior of the Billet Surface Layer in Porthole Die Extrusion of Aluminium, Metall. Mater. Trans., 31A, 1635-1643.

[12] Widerøe F., Welo T. (2012) Conditions for Sticking Friction between Aluminium alloy AA6060 and Tool Steel in Hot Forming, Key Eng. Mat., 491, 121-128.

[13] Sanabria V., Mueller S., Gall S., Reimers W. (2014) Investigation of Friction Boundary Conditions During Extrusion of Aluminium and Magnesium Alloys, Key Eng. Mat., 611-612, 997-1004.

[14] Sanabria V., Mueller S., Reimers W. (2014) Microstructure Evolution of Friction Boundary Layer During Extrusion of AA 6060, Proc. Eng., 81, 586-591.

[15] Jaspers S.P.F.C., Dautzenberg J.H. (2002) Material Behaviour in Metal Cutting: Strains, Strain Rates and Temperature in Chip of Formation, J. Mater. Process. Tech., 121, 123-135.

[16] Goldstein R.V., Alexandrov S.E. (2015) An Approach to Prediction of Microstructure Formation Near Friction Surfaces at Large Plastic Strains, Phys. Mesomech., 18, 223-227.

[17] Alexandrov S.E., Goldstein R.V. (2015) On Constructing Constitutive Equations in Metal Thin Layer near Friction Surfaces in Material Forming Processes, Dokl. Phys., 60, 39-41.

[18] Fries T.-P., Belytschko T. (2010) The Extended/Generalized Finite Element Method: An Overview of the Method and Its Applications, Int. J. Numer. Meth. Eng., 84, 253-304.

[19] Alexandrov S., Kuo C.-Y., Jeng Y.-R. (2015) A Numerical Method for Determining the Strain Rate Intensity Factor Under Plane Strain Conditions, Cont. Mech. Therm., 28, 977-992.

[20] Hwang Y.-M., Huang T.-H., Alexandrov S. (2015) Manufacture of Gradient Microstructures of Magnesium Alloys Using Two - Stage Extrusion Dies, Steel Res. Int. 86, 956-961.

[21] Alexandrov S., Jeng Y.-R., Hwang Y.-M. (2015) Generation of a Fine Grain Layer in the Vicinity of Frictional Interfaces in Direct Extrusion of AZ31 Alloy, Trans ASME J. Manuf. Sci. Eng. 137, Paper 051003.

[22] Alexandrov S. (2009) The Strain Rate Intensity Factor and Its Applications: A Review, Mater. Sci. Forum, 623, 1-20.

[23] Lyamina E., Nguyen T. (2016) Effect of the Yield Criterion on the Strain Rate and Plastic Work Rate Intensity Factors in Axisymmetric Flow, Struct. Eng. Mech. 58, 719-729.

[24] Druyanov B., Nepershin R. Problems of technological plasticity. Amsterdam: Elsevier; 1994.

[25] A. J. M. Spencer, "A theory of the failure of ductile materials reinforced by elastic fibres," Int. J. Mech. Sci., vol. 7, pp. 197-209, 1965.

[26] Hill, R., Lee, E.H., Tupper, S.J.: A method of numerical analysis of plastic flow in plane strain and its application to the compression of a ductile material between rough plates. Trans. ASME J. Appl. Mech. **18**, 46–52 (1951)

[27] Hill, R.: The Mathematical Theory of Plasticity. Clarendon Press, Oxford (1950)

The research on improving the anti-oxidation of tungsten rhenium alloy wires

Xinwei Bo[a,b],Qi Liu [a,b], Hao Yang [a,b],Chundong Jiang [a,b],Haoran He [a,b],Xiaoyu Wang [a,b],Chengchao Liu [a,b],Demao Chen [a,b]

dept. Refractory metal
[a]National Functional Instrument Materials Research Center,
[b]Chongqing Materials Research Institute Co., Ltd Chongqing, China

cqccy@qq.com

Abstract: The research compared the stability of EMF (Electromotive Force) though the same electrode testing method. Oxidation, morphology and composition of Tungsten Rhenium alloy thermocouple wire was characterized by Zeiss scanning electron microscope (EV018)and XPS (X-ray Photoelectron Spectroscopy) . The results showed that a small amount of rare earth elements droping tungsten rhenium alloy Wire can keep stable at a longer period of time in high temperature oxidation environment. Through the analysis of the SEM (Scanning Electric Microscopy) and XPS (X-ray Photoelectron Spectroscopy), the oxide film dopping rare earth is much denser, as well as lower valence oxide content. In a word, the rare earth droping in the tungsten rhenium alloy wire stabilize the EMF with specific films.

1. Introduction

Tungsten Rhenium thermocouple have been widely used in aviation, iron and steel smelting, petrochemical and other high-temperature test field because of its advantage in high melting point, high strength, high thermoelectric force, fast response speed and large signal output. Tungsten Rhenium thermocouple is quite susceptive at oxidizing atmosphere ,which restrict its range of application and storage method. Concerning on the effect of heat treatment on mechanical property[1] and failure analysis[2], the thermocouple of anti-oxidation generally include filament surface treatment and packaging technology with surface dip coating material[3] and package isolating oxygen[4]. Through the study of Tungsten Rhenium alloy wire oxidation behavior, the Kuihan Wang[5] and Ming Dai[6] found that the oxidation film grow radially under 800℃ at oxidation atmosphere. meanwhile, the J.R. Distefano[7] studied the oxidation behavior of Mo-Re alloy at different temperatures and pressures ,they found that the materials with larger oxygen solubility are brittle at low oxygen pressure. Our experiment research the anti-oxidant properties by adding Rare Earth Lanthanum elements in the Tungsten Rhenium alloy.

2. Experiment

2.1. Materials Preparation

In the test, the alloy with the composition of（97.5%～96.5%）W-（2.5%～3.5%）Re-（0.01%～0.1%）La in weight was prepared. To be noted that the purity of Tungsten powder - diameter 1～8μm

–is equal or greater than 99.9%.Rhenium added in form of Rhenium acid ammonia with purity not less than 99.99%; La added in form of analytical reagent Lanthanum Nitrate Hexahydrate . The preparation of W-Re-La alloy which have uniform composition and microstructure by metallurgic method containing weighing, high energy dispersion, low temperature roasting, high temperature deoxidize, adding forming agent, hydraulic pressure molding. Pre-sintering, sintering, density test, rotary forging, straight pulling, pull in the turntable and small turntable,500MM & 300MM fine drawing, electrolytic polishing and cleaning process, heat treatment.

2.2. Test method

The high precision digital multi-meter 8846A was used to measure and monitor the change of relative EMF (Electromotive Force) of the thermocouple wire which oxidize under 800℃; Using NETLSCH DIL449C thermo-gravimetric experiment equipment to test the oxidation process of WRe3 alloy (DSC/TG); The morphology and composition of oxidation film was characterized by Zeiss scanning electron microscope (EV018)and XPS (X-ray Photoelectron Spectroscopy) .

3. Analysis and discussion

3.1. The research of Rare Earth effect on EMF (Electromotive Force)

To improve the stability of thermocouple, the thermocouple should maintain the EMF (Electromotive Force) through standard index. The EMF (Electromotive Force) of WRe3 thermocouple wires （1#:La0.08%，2#:La0.1%）were tested in same polar method and found the non-uniformity is no more than 13μV (As shown in Table 1), which illustrate WRe3 alloy wires tolerance of temperature measurement under 0.125%.

Table1. The relative EMF of single Tungsten Rhenium thermocouple wire

Sample number	1#-1	1#-2	1#-3	2#-1	2#-2	2#-3
Relative EMF/μV	-4	3	5	1	-5	8

3.2. The stability testing of EMF in the air condition

Connecting two groups of K-type thermocouple wire with diameter 2.5mm as anode . In addition, the cathode separately used W-Re3 couple wire with diameter 0.5mm. The thermocouple wires were heated to 800±1℃ by electric heating tube furnace in the air. The FLUKE8846A digital multi-meter was used to measure the EMF of thermocouple, and record the data every 10 minutes, the results as shown in Figure 1.

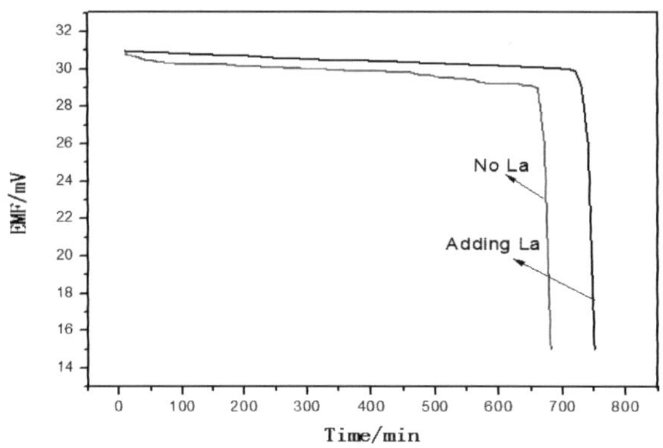

Figure 1. The EMF (Electromotive Force) of thermocouple at 800℃

The study found that the EMF (Electromotive Force) adding La perform more stable at 800℃ oxidation atmosphere, contrasting no La thermocouple wire. With time going on, EMF (Electromotive Force) sharply decrease, but the thermocouple doping La element works longer, just as shown in figure 1.

3.3 The analysis of oxidation process

Figure2 show the DSC/TG curves of W/Re alloy wires with Rare Earth La and without La element, respectively.

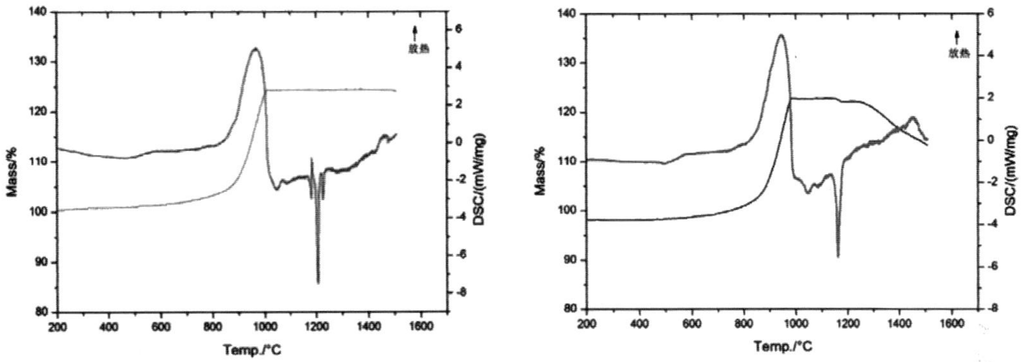

Figure 2. the DSC/TG contrast diagram of couple wires

The oxidation reaction of Tungsten Rhenium alloy in the air is an exothermic as well as height-increasing process, so the DSC/TG testing can be used to judge the oxidation . From the figure2, the Tungsten Rhenium thermocouple wire without doping La emerge a oxidation peak at 940℃,while doping La at 971℃.At the high temperature point, oxide film is easy to crack and peel off, which expose the wire to air and intensify oxidation reaction, thus leading to exothermic peak and weight increase significantly of wires. The DSC and TG curve peaks move to right by adding La element, which illustrate the importance of doping of Rare Earth La between oxide film and matrix enhanced.

3.4 The growth of oxide film

The micrograph morphology of W-Re3 alloy wires doping La or not oxidation at 850°Cwas showed in figure 3 and figure 4, respectively.

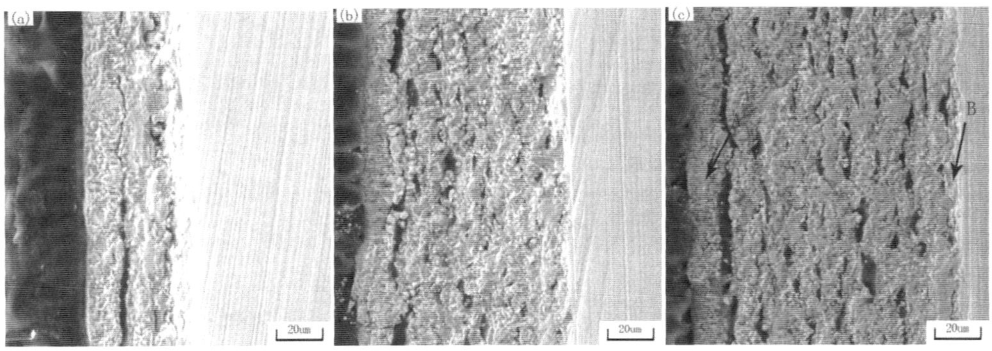

Figure 3. cross-section SEM images of WRe3 wires

（a）0.5h、（b）1h、(c) 1.5h

Figure 3 shows the oxidizing of WRe3 alloy wire without adding La at 850°C from 0.5h to 1h.With the extension of oxidation time , the film become more and more thick and more crack within oxidation film ,which caused large inner stress in the film and intensified the oxidation process. Fig.4 shows the oxidizing of WRe3 wires adding La under the same conditions, the Oxide film was compacted at 0.5h, even at 1.5h. No crack appears especially close to matrix, the permeate of oxygen to matrix/oxide film was hindered because of its compacted film. at the same time ,the film increase in a moderate way with time going on, which reflects that adding La in WRe3 thermocouple wire have more oxidation resistance than normal WRe3 thermocouple wire.

Figure 4. cross-section SEM images of WRe3 adding La
(a) 0.5h、 (b) 1h、(c) 1.5h

3.5. Analysis of oxide film composition

The scanning results of W are shown in Figure. 5 through XPS (X-ray Photoelectron Spectroscopy), there are two types of Tungsten oxides. First, Fig.5 (a) the main peak position 35.7eV, shoulder position 37.85eV, Genus WOx compound W6+ ion peaks characteristic. Corresponding to the 4f7 and 4f5 peaks of W6+, While the 41.5 eV peak belongs to the peak of W6+, corresponding to the 5p3 peak of W6+. The oxide type is believed to be WO3. Another category, The main peak position 35.9eV, shoulder position 38.05eV, The W5+ ion peaks of WOx ion compounds correspond to the 4f7 and 4f5 peaks of W5+. Fig.5 (b) are also two types of products, The main difference between the two membrane components is the W5+ content. Calculation shows that the content of W5+ in Lanthanum oxide film accounts for 19.33% of W element content, otherwise，W5+ element content accounts for 14.39% of W element content。The results showed that the oxidation resistance of Tungsten-Rhenium wire was improved to a certain extent.

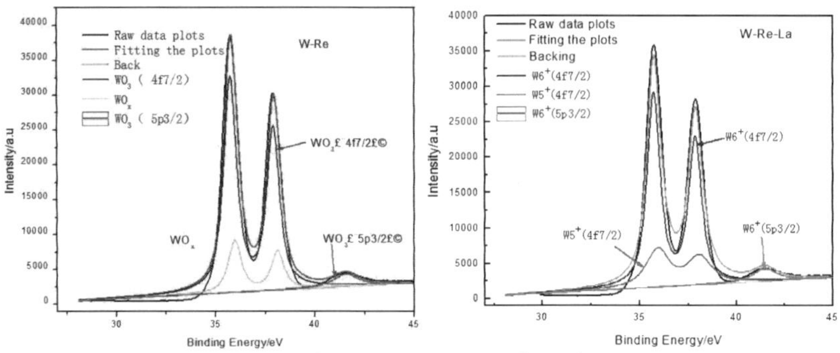

Figure 5. The XPS patterns of W element

4. Conclusion

(1)The EMF (Electromotive Force) is relatively stable within a period of time , before its significant mutations. Adding the Rare earth element in the wires, can prolong the stable state by 10% at 800 °C.

(2) By addition Rare earth ,The oxide film closing to the substrate is still relatively dense ,even oxidating at 850°Cfor 1.5 hours.

(3) Tungsten rhenium alloy wire can improve the anti-oxidation by droping rare earth element. which may be significant to enlarge the range of application and improve storage method of tungsten rhenium alloy wire, especially at low temperature oxidation atmosphere.

Acknowledgements

Financial support from Chongqing Foundation and Advanced Research Projects (Grant No. cstc2013jcyjA50027) and the Key Industry Common Generic Technology Innovation Projects （Grant No. cstc 2015zdcy-ztzx50005） is special gratefully acknowledged.

Reference

[1] Xinwei Bo and Qi Liu. The influence of Heat treatment process on the material mechanical properties for W-Re20 thermocouple wire [J]. Industrial Metrology, 2014, vol24, p.24-26.

[2] Hao Yang and Demao Chen. Failure analysis of Tungsten-Rhenium thermocouples [J]. Journal of Functional Materials . 2010， Proceedings: p.98-101.

[3] Zhongkui Yu and Yangzong Li. Molybdenum, oxidation coating for Tungsten and Tungsten Rhenium thermocouple wire [J]. Journal of South China University of Technology, 1984，vol,10，No1，: p.53-71.

[4] Suxia Zhang and Chuanbo Song. The manufacturing process and analysis of Anti-oxidant Tungsten-Rhenium thermocouples [J]. World Nonferrous Metals,2011, No.5: p.74-75.

[5] Kuihan Wang and Hongye He. High-temperature Stability of Tungsten-Rhenium thermocouples in air[J]. Acta Metrology sinica， 1997,vol18， No4： p.241-244.

[6] Min Dai and Kuihan Wang. Oxidation behaviors of W-5%Re/ W-26%Re Tungsten-Rhenium thermocouples.wires[J].Journal of Materials and Metallurgy,2014， vol 13， No1:p.55-59.

[7] J.R. DiStefano and B.A. Pint. Oxidation of refractory metals in air and low pressure oxygen gas [J]. International Journal of Refractory Metals & Hard Materials ,2000,vol 18): p. 237-243.

IOP Conf. Series: Journal of Physics: Conf. Series **885** (2017) 012020 doi:10.1088/1742-6596/885/1/012020

Laminar heat transfer with alumina nanofluid in cu tube with different diameters

Sedong Kim[a], A K M Mahmudul Haque[a], Junhyo Kim[b], Kitae Yu[c], Heegeun Song[c], Cheol Park[c], , Hanshik Chung[c], Hyomin Jeong[*c]

[a]Department of Energy and Mechanical Engineering, Gyeongsang National University, Tongyeong, South Korea; [b]Department of Marine Engineering, Mokpo National Maritime University, Cheonggye-myeon, South Korea; [c]Department of Energy and Mechanical Engineering, Institute of Marine Industry, Gyeongsang National University, Tongyeong, South Korea

hmjeong@gnu.ac.kr

Abstract. This study made nano particles-dispersing alumina nano fluid to three sized copper tubes in order to identify the differences in its heat transfer coefficient depending on its concentration. In order to evenly disperse nano particles in the fluid, it was treated with the ultrasonic process. And Thermal conductivities of nanofluid **are** measured via the LAMBDA measuring system by transient hot wire method. And to measure the differences in the nano fluid's dispersion degree depending on its different concentrations that was measured by mean of technique using a UV device. This study found that as the diameter of copper tube size was smaller, the heat transfer coefficient got higher and the fluid's heat transfer coefficient was changed depending on the different Reynolds numbers. Based on the findings of its experiment, this study explained the alumina nano fluid's characteristics in small diameter tubes

1. Introduction

A large number of industrial development and new technologies are limited by existing thermal management [1], and need for high performance cooling. A lot of industrial applications need ultrahigh-performance cooling systems to miniaturize the thermal systems. Nano fluid means the fluid having good thermal characteristics evenly dispersing nano particle on a fluid [2]. It is demonstrated that the advantages of nanofluids have better stability compared with those fluids containing micrometer or millimeter size particles and higher thermal conductive capability than their base fluids. thanks to the development of technology generating the nano particles, there have actively conducted several studies about the nano fluid[3-8] as a new form of heat-transferring fluid using the nano particles in traditional heat transfer fluids such as water, glycol, or oil base fluid [9–12] in recent decades. which was first propose by Dr. Stephen U.S.choi in Argonne National Lab, U.S.A in 1996[13]. And, Thermal conductivities of nanoparticle fluid mixtures have been reported by Eastman et al[14] and Artus[15] by adding a small volume fraction of metal or metal oxide powders in fluids increased the thermal conductivities of the particle-fluid mixtures over those of the base fluids. [16]Maxwell showed the possibility of increasing thermal conductivity of a mixture by greater volume fraction of solid particles. These fluids, containing colloidal suspended nanoparticles, have been called nanofluids. [17]General size of nano particle being used in a nano fluid is ranged from 10nm to 50nm, and the kinds of nano

IOP Conf. Series: Journal of Physics: Conf. Series **885** (2017) 012020 doi:10.1088/1742-6596/885/1/012020

particles having been currently used are the Al2O3, CuO, Cu, Pt, Au single-layer structured nano tube and the multiple-layer structured nano tube. This study aimed to produce various concentrations of nano fluid in use of alumina nano particles among various types of nano particles and to identify the pure water' and the alumina nano fluid's thermal characteristics through a basic heat-transfer experiment depending on different tube sizes.

2. Experimental details

In this experiment, Distilled water (DW) is used for the base fluid for nanofluid in present experiment. the researcher made the nano fluid having the concentrations of 0.5wt%, 1.0wt% and 2.0wt% in order to identify the alumina nano fluid's heat transfer characteristics. Figure 1 showed the distilled water and three nano fluids having different concentrations which were really produced,

Figure 1. The Image of DW and Al2O3 (0.5wt%,1.0wt% and 2.0wt%)

2.1. Experiment Method

In this experiment, the researcher composed the following experiment tool as shown in the Figure 2 in order to identify the distilled water's and alumina nano fluid's heat transfer coefficient characteristics

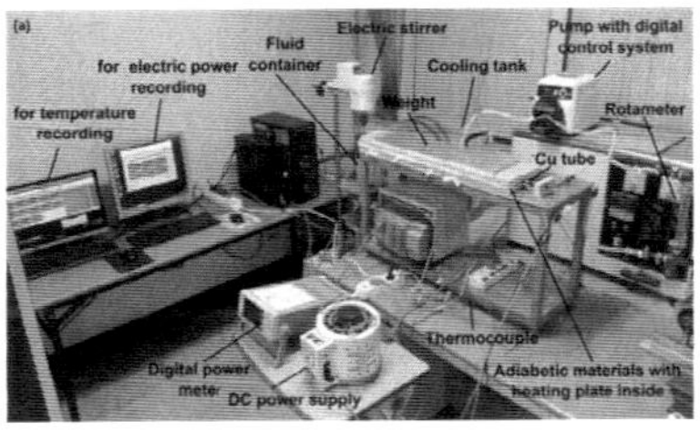

Figure 2. (a) The digtal Photo of experimental apparatus.

First, for the test section to find out the both materials' heat transfer coefficient characteristics, the investigator first prepared a heating plate having its length=0.76m and width=0.12, and experimented with three concentrations of 0.5wt%, 1wt% and 2wt% alumina fluids and distilled water being contained in 0.8mm, 1.6mm and 2mm and 1m-long copper tubes. Outside of copper tubes in the test section, some T-type thermocouples were attached at intervals of 0.152m, and the thermocouples were wrapped with some insulating materials. Concerning the fluids flow, the investigator used pump to supply fluid in a certain flow rate. And the fluid being cooled down in a water tank was transferred by pump to the copper tube and then the fluid was cooled down in a tank again. Like that the fluid was gone through the repetitive process. For the measuring instruments, this experiment used the data logger for storing each fluid's temperature data and the electricity power-meter in order to measure the input calorie. This experiment compared the results of distilled water, 0.5wt%, 1wt% and 2wt% in order to prove this experiment's results. Performance of local heat transfer in terms of convective heat transfer coefficient is calculated by,

$$H(x)=\frac{Q}{A\,(T_w-T_f)_x} = \frac{Q}{lw\,(T_w-T_f)_x}$$

$$= \frac{\ddot{q}}{(T_w-T_f)_x}$$

So, the actual heat flux can be calculated by the equation

$$\ddot{q}=\frac{Q}{\pi dL}=\frac{Q}{lw}$$

Fluid temperature profile in the test section can be calculated as,

$$T_f=T_{in}+\frac{\ddot{q}Px}{\rho C_p v A}$$

3. Result and discussion

In order to evenly disperse nano particles in the fluid, it was treated with the ultrasonic process. To characterize the dispersion of the Nanofluid suspensions UV measurement system was measured and compared with each date. Fig 3 shows the UV spectra of the distilled water and nanofluid (0.5wt%, 1.0wt%and 2.0wt%). As shown, the lowest absorbance was measured for the 0.5wt% suspension except distilled water. Meanwhile, the 2.0wt% suspension exhibited much higher absorbance than 0.5wt%, 1.0wt%

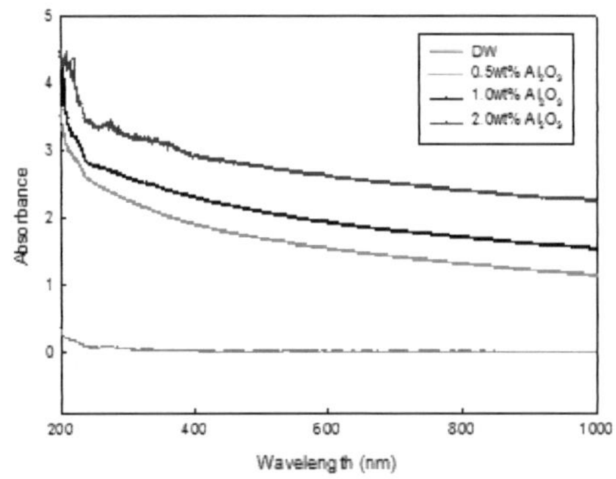

Figure 3. Absorbance test comparing different concentration nano fluids

Figure 4 showed the results comparing different diameter of copper tube depending on different alumina concentrations in each Reynolds number. As the concentration of alumina nano fluid got increased, the heat transfer coefficient tended to be increased, and it could be seen that smaller diameter of copper tube had higher heat transfer coefficient than bigger diameter of copper tube. And it could be seen that as the number of Reynolds was increased, the gap of heat transfer coefficient each tube , and the heat transfer coefficient was increased.

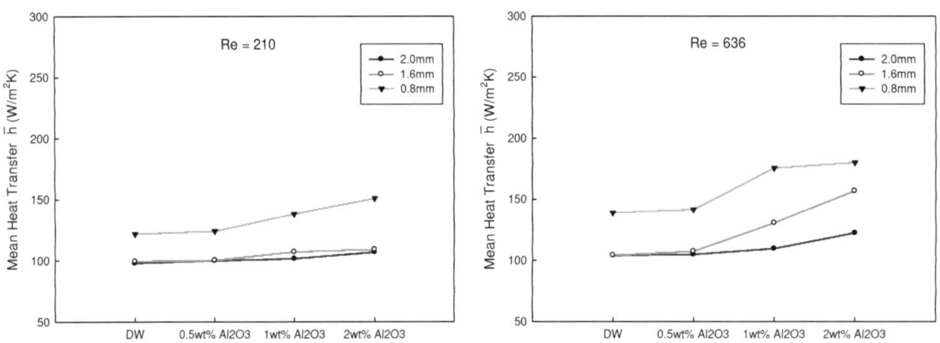

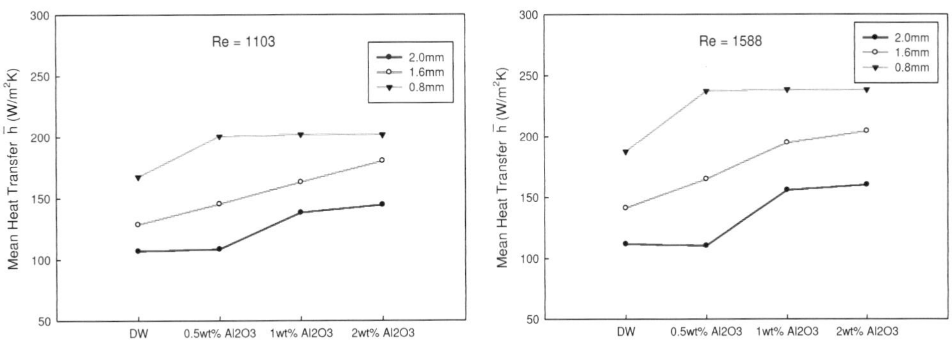

Figure 4. Experimental date for each Reynolds number.

4. Conclusion

This experiment conducted a heat transfer experiment in using the Distilled Water, 0.5wt%, 1.0wt% and 2.0wt% concentration of nano fluid in order to identify alumina nano fluid's heat transfer characteristics, and got the following conclusion.

From this experiment, it was identified that as the concentration of alumina nano fluid was increased, the fluid's heat transfer coefficient was higher, and as the diameter size of tube was smaller, the tube's heat transfer coefficient got higher. Additionally, for the case that the number of Reynolds was 210, the 0.8mm diameter of copper tube's heat transfer coefficient was increased to 131.64% than that of 2mm diameter of copper tube, and for the case that the number of Reynolds was 636, the 0.8mm diameter of copper tube's heat transfer coefficient was increased to 144.33% than that of 2mm diameter of copper tube. And for the case that the number of Reynolds was 1103, the 0.8mm diameter of copper tube's coefficient was increased to 154.899%, and for the case that the number of Reynolds was 1588, the 0.8mm diameter of copper tube's coefficient was increased to 167.53%. Judging from these results, it could be also identified that as the number of Reynolds was increased, the increase margin of heat transfer coefficient was increased according to that.

Nomenclatures

A: Cross-sectional area, [m2]
CP: Specific heat, [J/Kg K]
D: Tube diameter, [m]
DW : Distilled water
H : Heat transfer coefficient, [W/m2K]
K: Thermal conductivity, [W/m K]
L: Length of the test section, [m]
P: Perimeter [m]
Q: Watt [W]
$\ddot{q}$: Heat flux, [W/m2]
w : Width [m]
Re: Reynolds number
T : Temperature, [K]
v: Velocity, [m/s]
vol.%: Volume percentage
wt.%: Weight percentage
X: axial distance from the tube entrance [m]
v: Dynamic viscosity (kg/m.s)

C_p: Specific heat (J/kgK)
ρ: Density (kg/m3)
δ: Boundary layer thickness (mm)
f :fluid
in: inlet
out:outlet
w:tube wall

Acknowledgment
This research was supported by Basic Science Research Program through the National Research Foundation of Korea (NRF) funded by the Ministry of Science, ICT and future Planning (No. 2017R1AB4007620).

References

[1] Hafizur Rehman,"Experimental and Numerical Analysis of Convective Heat Transfer to Alumina Nanofluids Under Laminar Flow Regime." Journal of Computational and theoretical Nanoscience vol.10, 2305-2311,2013
[2] Y. G. Kim "Experimental investigation of heat transfer characteristics of alumina nanofluid" , Journal of the Korean Society of Marine Engineering, vol. 37 no.1 pp. 16~21, 2013.
[3] Lee, S, Choi, S. U. S. and Eastman, J.A.,1999, Measuring thermal conductivity of fluids containing oxide nanoparticles, ASME J. Heat Transfer, Vol,121, pp. 280~289.
[4] Eastman, J.A., Choi, S.U.S., Yu, W, and Thompson, L., J.,2001, Anomalously increased effective thermal conductivity of ethylene glycol-based nanofluids containing copper nanoparticles, Appl. Phys. Lett, Vol.78, pp.718-720.
[5] Choi, S.U.S., Zhang, Z.G., Yu, W, Lockwood, F.E. and Grulke, E.A., 2001, Anomalous thermal conductivity enhancement in nanotube suspensions. Appl. Phys. Lett., Vol.79, pp.2252-2254
[6] Das, S.K., Putra, N.,Thiesem.P.and Roetzel, W.,2003, Thermal conductivity of naked and monolayer protected metal nanoparticle base nanofluids: Manifestation of anomalous enhancement and chemical effects, Appl. Phys. Lett.,Vol.83,pp.2931-2933.
[7] Jang, S.P. and Choi, S.U.S., 2004, Role of Brownian motion in the enhanced thermal conductivity of nanofluids, Appl. Phys. Lett., Vol.84, pp.4316-4318.
[8] You, S.M., Kim, J.H. and Kim, K.H., 2003, Effect of nanoparticles on critical heat flux of water in pool boiling heat transfer, Appl. Phys. Lett., Vol.83, pp.3374-3376.
[9] X.F. Li, D.S. Zhu, X.J. Wang, J.W. Gao, H. Li, Thermal conductivity enhancement dependent pH and chemical surfactant for Cu-H2O nanofluids, Thermochimica Acta 469 (1–2) (2008) 98–103.
[10] J.A. Eastman, U.S. Choi, S. Li, L.J. Thompson, S. Lee, Enhanced Thermal Conductivity through the Development of Nanofluids, Fall Meeting of the Materials Research Society, Boston, Dec 2–6, 1996, in: Materials Research Society, Pittsburgh, 1996, pp. 1–12.
[11] E.V. Timofeeva, A.N. Gavrilov, J.M. McCloskey, Y.V. Tolmachev, et al., Thermal conductivity and particle agglomeration in alumina nanofluids: experiment and theory, Physical Review E 76 (2007) 061203–061220.
[12] J.A. Eastman, S.R. Phillpot, S.U.S. Choi, P. Keblinski, Thermal transport in nanofluids, Annual Review Material Research 34 (6) (2004) 219–246.
[13] S. Choi, U. S. Enhancing thermal conductivity of fluids with nanoparticles, in:
[14] Proceedings of the 1995 ASME International Mechanical Engineering Congress and Exposition, San Francisco, CA, Nov 12-17,1995, vol. 231, American Society of Mechanical Engineers, New York, 1995, pp. 99–105.
[15] J. A. Eastman, U. S. Choi, S. Li, L. J. Thompson, and S. Lee, Enhanced thermal conductivity through the development of nanofluids, Materials Research Society Symposium Proceedings, Materials Research Society, Pittsburgh, PA (1997), Vol. 457 pp. 3.11.
[16] G. R. C. Artus, IEEE Transactions on Components, Packaging, and Manufacturing—Part B 19,

601 (1996).
[17] J. C. Maxwell, A Treatise on Electricity and Magnetism, Oxford University Press, Cambridge (1881).
[18] S. P. Jang "Cooling Performance of a Microchannel Heat Sick with Nanofluids", Journal of School of aerospace and Mechanical Engineering, vol.17 no.9 pp. 849~854, 2005.
[19] J.P. Bently, Temperature sensor characteristics and measurement system design, Journal of Physics E: Scientific Instruments 17 (6) (1984) 430–435

IOP Conf. Series: Journal of Physics: Conf. Series **885** (2017) 012021 doi:10.1088/1742-6596/885/1/012021

CFD Analysis of nanofluid forced convection heat transport in laminar flow through a compact pipe

Kitae Yu[1], Cheol Park[1], Sedon Kim[1], Heegun Song[1], Hyomin Jeong[2]

[1]Department of Energy Mechanical Engineering Gyeongsang National University Cheondaegukchi-Gil 38, Tongyeong, Gyeongnam, 650-160, Republic of Korea
[2]Department of Energy and Mechanical Engineering Gyeongsang National University Institute of Marine Industry
Cheondaegukchi-Gil 38, Tongyeong, Gyeongnam, 650-160, Republic of Korea

hmjeong@gnu.ac.kr; belljidanto@gmail.com; Lu_sky90@daum.net; opoq123@naver.com; Gmlrms1016@gmail.com

Abstract. In the present paper, developing laminar forced convection flows were numerically investigated by using water-Al2O3 nano-fluid through a circular compact pipe which has 4.5mm diameter. Each model has a steady state and uniform heat flux (UHF) at the wall. The whole numerical experiments were processed under the Re = 1050 and the nano-fluid models were made by the Alumina volume fraction. A single-phase fluid models were defined through nano-fluid physical and thermal properties calculations, Two-phase model(mixture granular model) were processed in 100nm diameter. The results show that Nusselt number and heat transfer rate are improved as the Al2O3 volume fraction increased. All of the numerical flow simulations are processed by the FLUENT. The results show the increment of thermal transfer from the volume fraction concentration.

1. Introduction

To enhance the heat exchangers energy transfer efficiency, Engineers have interested not only the development exchanger or enhanced its structure, but also improvement of heat transfer fluid itself. Maxwell [1] proved the possibility of increasing thermal conductivity of a mixture by higher volume fraction of solid particle. Nowadays available particle diameter is smaller than 100nm which is called 'Nano-sized'. This effort makes particle mixture fluid would enhance the heat transfer performance of heat transfer fluids. [2] Moreover, solid nanoparticle colloids are highly stable and exhibit no significant settling under static conditions. Also, decreasing particle size makes suspending solid colloid easily. So that nano-fluid technology is expected to suitable of heat transfer fluid. [3] However, nano-fluid development is interrupted by the lack of agreement with different research group, theoretical understanding of the heat transfer mechanisms, different suspension conditions. [4] In this paper, Nano-fluid model assume that the Homogeneous flow mixture model [5] which is the dispersed and continuous phase with very strong coupling and the nano-particles is moving at the same inlet velocity. Therefore instead of DPM, the dispersed granular mixture models are made by 'Mixture model' which is one of the Euler-Euler approaches. This numerical investigation purpose is an observation of heat transfer coefficient increment in a compact pipe and comparison between single-phase and multi-phase model. Geometry model size is as same as experimental pipe which will use in a university laboratory. Nano-particle volume fractions are less than 5 percent, and divided by 1 percent from 0 to 4 percent. The wall has constant surface heat flux as known as Uniform Heat Flux (UHF). The numerical simulation geometry is axisymmetric two-dimensional model. Through the limit of 2D model and assumption, implicit effect such as gravity and buoyancy force are excluded. The nano particle material is an aluminum oxide with a spherical size of 100 nm diameter and simulations are processed in a steady state.

2. Mathematical modelling

2.1. Geometry and boundary condition

Fig. 1a shows the actual model of geometry and Fig. 1b is the part of the Two-dimensional Mesh. One of the main purpose of this simulation is trying to understand the near wall behavior of nano-particles, velocity distribution perpendicular to the pipe surface and drag coefficient near at the pipe wall, so the relatively concentrated mesh in the near wall area is needed. As a result, the mesh has a non-uniform quadratic mesh. The bottom horizontal line is an axis which makes Three-dimensional pipe model at the axisymmetric environment. The upper horizontal line is a wall with UHF. The tube consisted with a diameter (D) of 4.5 mm and a length (L) of 1.17 m. The nano-fluid is composed of water and Al_2O_3 particles. Whole velocity inlet for each fluid models enter with uniform temperature and axial velocity profiles are defined by Re = 1050. The fluid is flow out to the outflow outlet.

2.2. Governing equation

The single-phase model is investigated for the comparison of mixture model as prescribed. Following equations are representing the mathematical formulation of the single-phase model and of the continuous phase of the mixture model. [5-7]

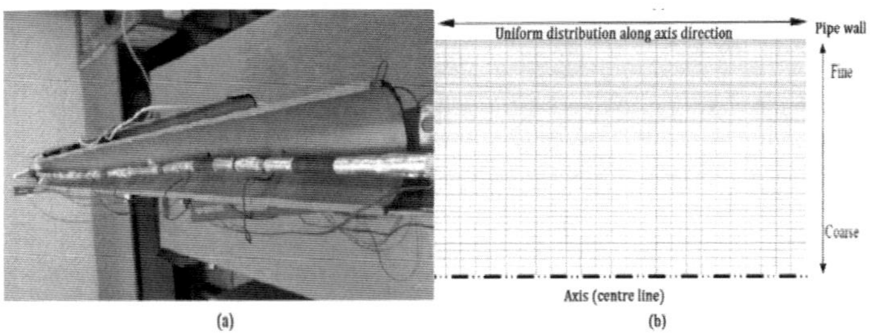

(a) (b)

Figure 1. (a) Actual model of geometry (b) Part of the mesh of Computational Cell. 80 x 8000 with 0.96 ratio

Energy equation

$$div(\rho\vec{V}CpT) = div(kgradT) + S_e \tag{1}$$

Conservation of mass

$$\frac{\delta\rho_m}{\delta t} + \nabla\cdot(\rho_m u_m) = 0 \tag{2}$$

Mass-averaged velocity u_m is

$$u_m = \frac{\sum_{k=1}^{n}\alpha_k\rho_k u_k}{\rho_m} \tag{3}$$

and ρ_m is the mixture density with α_k volume fraction of phase *k*

$$\rho_m = \sum_{k=1}^{n}\alpha_k\rho_k \tag{4}$$

Momentum equation for single-phase

$$div(\rho\vec{V}\vec{V}) = -grad\,P + \nabla\cdot(\mu\nabla\vec{V}) + S_m \tag{5}$$

Momentum equation for multi-phase

$$\nabla\cdot\sum_{k=1}^{n}\alpha_k\rho_k u_k u_k = \nabla\cdot(\rho_m u_m u_m) + \nabla\cdot\sum_{k=1}^{n}\alpha_k\rho_k u_{Mk}u_{Mk} \tag{6}$$

The compression and viscous lose by heat are assumed to be excluded in the energy equation. The source/sink terms S_m and S_e represent the unitive effects of energy and momentum exchange with base fluid. These are zero value in the single-phase model. Momentum equation of multi-phase are affected by volume fraction (α), velocity of phase(u_k), diffusion velocity (u_{Mk}). The determinations of particle motion are particle density(ρ_c), cross-sectional area of the particle(A_p), drag coefficient(C_D), slip velocity(u_{cp}). The drag force in this study is following Clift et al equation (7) for single rigid spherical particle in a fluid

$$F_D = -\frac{1}{2}A_p\rho_c C_D\left|u_{cp}\right|u_{cp} - \frac{1}{2}V_p\rho_c\frac{du_{cp}}{dt} - 6\gamma_p^2\sqrt{\pi\rho_c\mu_c}\int_0^t \frac{\frac{du_{cp}}{ds}}{\sqrt{t-s}}ds \tag{7}$$

The drag coefficient which is already showed in equation (7) depends on small Particle Reynolds numbers with particle Reynolds number

$$C_{D,St} = \frac{24}{Re_p} \tag{8}$$

The slip velocity is following equation (9) with τ_p the particle relaxation time, $\vec{a}$ is the secondary-phase particle's acceleration

$$u_{pq} = \frac{\tau_p(\rho_p - \rho_m)}{f_{drag}}\vec{a} \tag{9}$$

The default drag function is taken Schiller and Naumann. Reynolds numbers in this paper are over 1000 so the equation is defined as follow

$$f_{drag} = 0.0183\text{Re} \tag{10}$$

2.3. *Physical properties of the nanofluid for single-phase model and Boundary conditions*

At inlet boundary condition, axial velocity V_0 is defined from $Re = 1050$ according to the fluid properties and a constant temperature (293K) as described. On the tube wall, the usual non-slip conditions and uniform wall heat flux $q = 5000\ W/m^2$ are implemented.

The determination of nano-fluid physical, thermal properties are at the center of current nano-fluids research as previously mentioned.[7] For the single phase model properties of the nano-fluids have been calculated by using classical formulas developed for conventional solid-liquid mixtures.[8] The following equations are considered to evaluate Al_2O_3-water nano-fluids thermo-physical properties and were used for CFD simulation.[6]

$$\rho_{nf} = (1-\varphi)\rho_{bf} + \varphi\rho_p \tag{11}$$

$$c_{p,nf} = \frac{(1-\varphi)(\rho c_p)_{bf} + \varphi(\rho c_p)_p}{\rho_{nf}} \tag{12}$$

$$\mu_{nf} = (123\varphi^2 + 7.3\varphi + 1)\mu_{bf} \tag{13}$$

The φ is the volume fraction of the Al_2O_3. The dynamic viscosity that temperature depended properties are excluded and has a constant value. The equation (13) is applicable for spherical particle in volume fractions less than 5% [9] and determination of nano-fluid effective thermal conductivity as follow. [10]

$$k_{nf} = \left[\frac{k_s + 2k_w + 2(k_s - k_w)(1+\beta)^3\varphi}{k_s + 2k_w - (k_s - k_w)(1+\beta)^3\varphi}\right]k_w \tag{14}$$

In Equation (14), β is the ratio of nano-layer thickness to the original particle radius, and $\beta = 0.1$ is used for calculation in this equation. The result of physical and thermal properties of nano-fluid models are presented in Table 1 and Al_2O_3 particle physical and thermal properties as follow in Table 2

2.4. *Numerical approach*

The computational fluid dynamic code FLUENT 16.1 is employed to solve the present problem. The governing Eqs. (1)-(3) are solved by control volume approach. For the convective and diffusive terms, a second order upwind method was used. Pressure and velocity were coupled using Semi Implicit Method for Pressure Linked

Table1. Properties of nano-fluid at 293K and Re = 1050

Φ(%)	*Density (kg/m^3)*	*Viscosity (kg/ms)*	*Inlet velocity (m/s)*	*Heat conductivity (W/mK)*	*Specific heat (J/kgK)*
0	*998.2*	*9.98e-4*	*0.233*	*0.5970*	*4182.0*
1	*1027.018*	*1.023e-3*	*0.2324*	*0.6200*	*4053.21*
2	*1055.836*	*1.048e-3*	*0.2316*	*0.6436*	*3931.451*
3	*1084.654*	*1.073e-3*	*0.2308*	*0.6678*	*3816.162*
4	*1113.472*	*1.098e-3*	*0.2300*	*0.6926*	*3706.840*

Table 2. Al_2O_3 nanoparticle at temperature of 293K

Chemical formula	*Density (kg/m^3)*	*Heat conductivity (W/mK)*	*Specific heat (J/kgK)*	*Spherical diameter (nm)*
Al_2O_3	*3880*	*36*	*773*	*100*

Equations (SIMPLE) in single-phase. Unlike single-phase, multi-phase was approached by Coupled algorithm which offers an alternative to the density-based and pressure-based segregated algorithm with SIMPLE-type pressure-velocity coupling. The difference between SIMPLE and Coupled is a Convergence time [6] and Couple algorithm has converged faster than SIMPLE in multi-phase.
The laminar flow has fully developed region that distribution of flow velocity maintains constant. In this study, Fully developed region appears after 0.307m according to equation (15). [12].

$$L_{laminar} = 0.065 \cdot Re \cdot D \tag{15}$$

The velocity distributions are showed in Figure 2. It is only 0.6737 percent less than outlet at an axial velocity at the 0.307m in Fig. 2a and all mixture models are following the Fully developed region in Fig. 2b. This means nano-fluid motion is fully developed at exit section model, when properties are not dependent on temperature, velocity profile and hydrodynamic entrance length are nearly identical at 0.307m for Φ = 0%, 1%, 2%, 3% and 4%.[7]

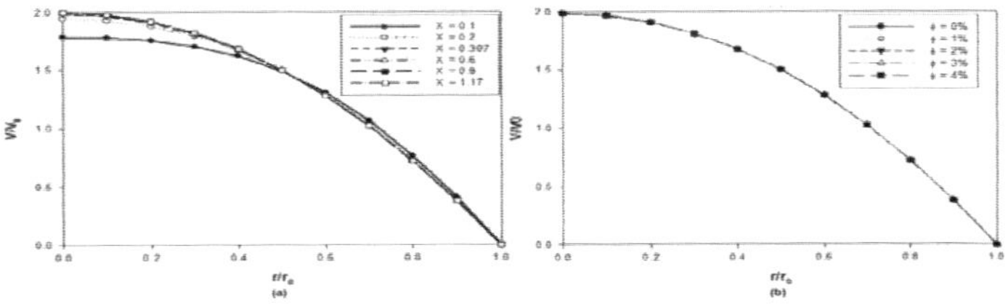

Figure 2. (a) Profiles of axial velocity at several locations for Re = 1050, q = 5000 W/m^2, Φ = 4%. (b) Profiles of axial velocity at x = 0.307m for Re = 1050, q = 5000W/m^2 and several volume fractions.

A Nusselt number is converged to 4.36 in the UHF.[13] Comparing Nusselt number to Churchill-Ozoe equation (16) [14] and Siegel Sparrow-Hallman calculation equation (17) [15] and common Nusselt number equation (18) which used in present study in Figure 3. x^* is a dimensionless axial position for thermally developing flows. Pr is the Prandtl number.

$$\mathrm{Nu} = 5.364\left[1 + (70.0x^*)^{-10/9}\right]^{3/10} - 1 \tag{16}$$

$$Nu = \frac{2}{\left[\left(\frac{t_w - t_0}{qr_0/k}\right) - \left(\frac{4x/r_0}{RePr}\right)\right]} \tag{17}$$

$$Nu = \frac{h(x)D}{k_{inlet}} \tag{18}$$

The heat transfer coefficient h(x) is defined as

$$\mathrm{h}(x) = \frac{q}{T_w(x) - T_b(x)} \tag{19}$$

The $T(x)_w$ is the solid surface temperature at the wall temperature, and $T(x)_b$ is the mean bulk temperature. Average Nusselt number is defined as

$$\mathrm{Nu}_{av} = \frac{h_{av} \cdot D}{k_{inlet}} \tag{20}$$

Current study model on $\varphi = 0\%$ is going to 4.69 Nusselt number which has a 7 percentage point gap between normal convergence value. Therefore velocity distribution and Nusselt number show the model has fitted in the theory.

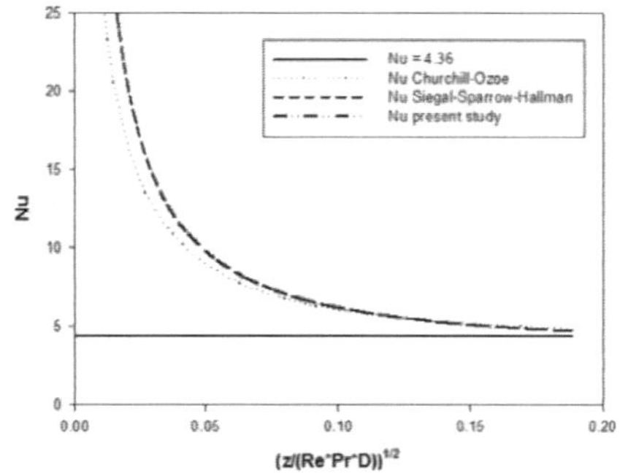

Figure 3. Grid validation by means of Churchill and Ozoe, Siegal-Sparrow-Hallman

3. Results

In this work a number of numerical simulations have been performed to study forced convective heat transfer of Al_2O_3-water in a circular tube under laminar flow and results were conducted to employing the single phase and mixture model for $\boldsymbol{\varphi} = \mathbf{1}\%$ to $\mathbf{4}\%$, $\mathbf{Re} = \mathbf{1050}$ as $\mathbf{q} = \mathbf{5000}\, \boldsymbol{W/m^2}$ with a constant state. In all cases the size of the spherical particles is 100 nm. Dimensionless temperature shows the particle concentration increase happens the thermal development acceleration of the fluid at several axial locations along the radius is reported in Fig. 4 for $\boldsymbol{\varphi} = \mathbf{0}\%$ and $\mathbf{4}\%$. For the both the base fluid and the nano-fluid, the motion is not thermally developed. As it is also noticed by a higher slope of relative local heat transfer coefficient h_r for $\boldsymbol{\varphi} = \mathbf{4}\%$ shown in Fig. 5a. Relative local heat transfer coefficient is increasing with axis location and volume fraction rise when fluid properties are constant in Fig. 5a. Because of narrow hydrodynamic radius of model, correlation between wall and bulk temperature is higher than larger model. Consequently, difference between the wall and bulk temperature is decrease and with the heat flux on the wall being constant, there is an increase in the heat transfer coefficient, as shown in Fig. 5b.Small displayed equations: Some examples:

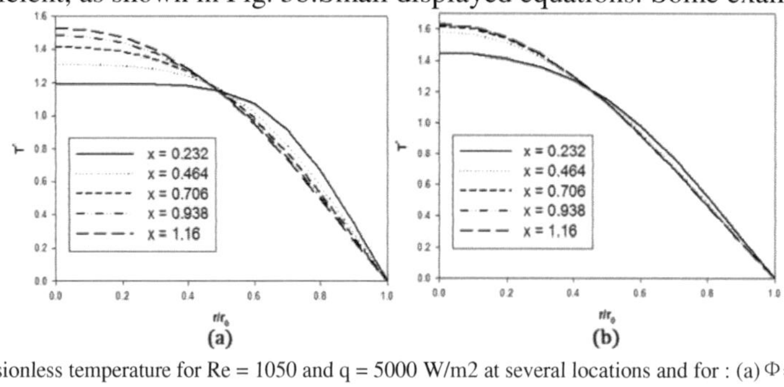

Figure 4. Dimensionless temperature for Re = 1050 and q = 5000 W/m2 at several locations and for : (a) Φ =0% and (b) Φ = 4%

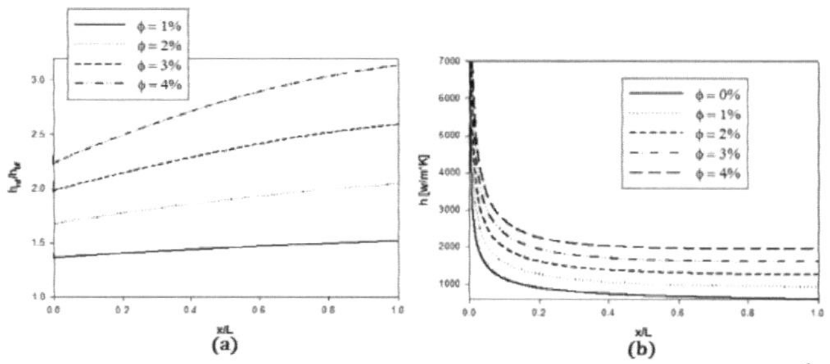

Figure 5. (a) Increase in nano-fluid heat transfer coefficient along tube axis for Re = 1050 and q = 5000 W/m^2 for constant properties and increment of volume fraction. (b) Heat transfer coefficient of each volume fraction for constant properties.

In Fig. 6, radial temperature profiles at profiles at x/L = 1.0 for Re = 1050 and q = 5000 W/m^2 are investigated for the effect of particle presence. The temperature is strongly dependent on φ

IOP Conf. Series: Journal of Physics: Conf. Series **885** (2017) 012021 doi:10.1088/1742-6596/885/1/012021

concentration. When the multi-phase concentration increases, the difference between temperature at surface and axis decrease by higher forced convection heat transfer. This effect is the cause of wall temperature decrease. It is also implemented in single-phase however, due to the particle absence, heat transfer between wall and inner side is not remarkable. Furthermore the only average temperature 1.3K increase than 293K on single-phase models. Both of phase show there is a clear existence of a uniform temperature fluid zone that becomes more visible for higher value of Φ at the $r/r_0 = 0.6$.

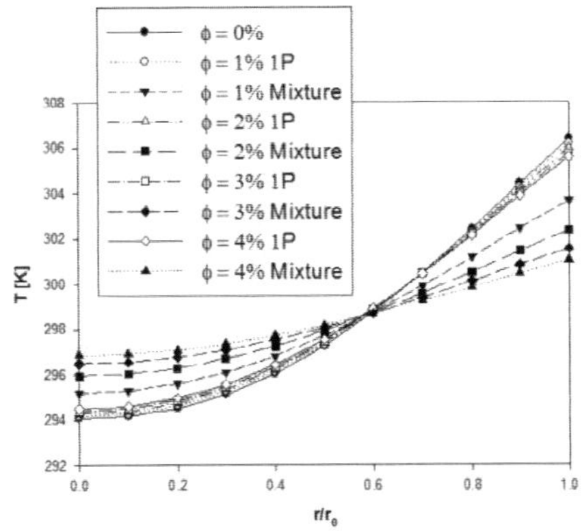

Figure 6. Comparison of temperature along tube radius at x/L = 0.99 for several concentration and for Re = 1050 and q = 5000 W/m^2 in constant properties

Table3. Heat transfer coefficients for Re = 1050, q = 5000 W/m^2 and the investigated models

Φ(%)	h_{av} (W/mK)	h_{nf}/h_{bf}	Nu_{av}	Nu_{nf}/Nu_{bf}
0	861.815	1	6.5	1
1	1233.288	1.431	8.95	1.377
2	1592.853	1.848	11.1	1.707
3	1954.472	2.268	13.2	2.027
4	2301.521	2.671	14.9	2.302

4. Conclusion

Average heat transfer coefficient and Nusselt number are shown in Table 3 with relative increase of the total heat transfer rates and Nusselt number as a concentration of the nano-particle volume fraction φ. Significant increases of the total heat transfer rates can be found with the use of fully dispersed nanoparticles multi-phase. These results have indicated the nano-fluids beneficial effects of thermal properties improvement. The results showed that the nanoparticles inclusion produced considerable heat transfer with respect to base fluid. Heat transfer enhancement was increasing with concentration of particle volume. But the CFD should be progressed with constant and transient numerical investigation parallel and these numerical researches would be pace with the experiment especially. Also, lack of Reynolds number and uniform heat flux example, reality of nano-particle dispersed state are not including in this paper.

Acknowledgments

This research was supported by Basic Science Research Program through the National Research Foundation of Korea(NRF) funded by the Ministry of Education(No. 2015R1D1A1A01058030).

References

[1] J. C. Maxwell, A Treatise on Electricity and Magnetism, Oxford University Press, Cambridge, 1881

[2] Taylor, R. A, Small particles, big impacts: A review of the diverse applications of nanofluid, Journal of Applied Physics 113, pp. 1, 2013

[3] S. M. S. Murshed, K. C. Leong, C. Yang, Thermophysical and electrokinetic properties of nanofluids – a critical review, Applied Thermal Engineering 28, pp. 2109-2125, 2008

[4] X. Q. Wang, A. S. Mujumdar, Heat transfer characteristics of nanofluids: a review, International

Journal of Thermal Sciences 46, pp. 1-19, 2007

[5] Mi. Manninen, V. Taivassalo, S. Kalliio, On the mixtrue model for multiphase flow, Valtiion Teknillinen Tutkmuskeskus, 1996

[6] ANSYS Fluent Theroy Guide, pp. 501, November 2013

[7] V. Bianco, F. Chiacchio, O. Manca, S. Nardini, Numerical investigation of nanofluids forced convection in circular tubes, Applied Thermal Engineerinig 29, pp. 3632-3642, 2009

[8] S. K. Das, S. U. S. Choi, W. Yu, and T. Pradeep, Nanofluids Science and Technology, John Wiley & Sons, Hoboken, Nj, USA, 208

[9] D. A. Drew, S. L. Passman, Theory of Multicomponent Fluids, Springer, Berlin, 1999

[10] S. J. Kline and F. A. McClintock, Mechanical Eng. 75, 3, 1953

[11] M. Nazififard, M. Nematollahi, K.Jafarpur, K. Y. Suh, Numerical Simulation of Water-Based Alumina Nanofluid in Subchannel Geometry, Science and Technology of Nuclear Installations, Vol 2012, 2012

[12] Kominu Natsuo, Fluid Engineeriing, Syuwa system, pp. 176, 2006

[13] Mohammad Z. Hasan, Effects of non-uniform surface heat flux and uniform volumetric heating on blanket design for fusion reactors, Fusion Science and Technology, Los Angeles, May 1988

[14] S. W. Churchill, H. Ozoe, Correlations for forced convection with uniform heating in flow over a plate and in develooping and fully developed flow in a tube, Journal of heatTrnasfer 95, pp. 78-84, 1973

[15] R. Siegel, E. M. Sparrow, T. M. Hallman, Steady laminar heat transfer in a circular tube with prescibed wall heat flux, Cleveland, Ohio, UA, Aplied Scientific Research, Section A, Volume 7, Issue 5, pp. 386-692, September, 1958

IOP Conf. Series: Journal of Physics: Conf. Series **885** (2017) 012022 doi:10.1088/1742-6596/885/1/012022

Application of $CaCO_3$, CaF_2, SiO_2, and TiO_2 particles to silicone lens for enhancing angular color uniformity of white LED lamp

Anh Q. D. Nguyen, Vinh H. Nguyen

Nguyen Tat Thanh University, Ho Chi Minh City, Vietnam

anhkuas@gmail.com

Abstract. In this study, we present an insightful investigation on optimal selection of scattering enhancement particles (SEP) to satisfy each specific optical property of white LEDs (WLEDs). The interested contenders include $CaCO_3$, CaF_2, SiO_2, and TiO_2, each of them is added with YAG:Ce phosphor compounding. The quality improvement on each considered property is demonstrated convincingly by applying Mie-scattering theory together with Monte Carlo simulation on a particular WLEDs which has the color temperature of 8500K. It is observed by simulation results that TiO_2 particles provide the highest color uniformity among the SEP, as increasing TiO_2 concentration. These results of this work can serve as a practical guideline for manufacturing high-quality WLEDs.

1. Introduction

To improve the overall performance of WLEDs, previous studies have focused on three important optical properties, including color uniformity, luminous flux, and color rendering ability [1]. All these authors' approaches met at one point, which is to enhance scattering events in phosphor-converted white-LEDs (pcLEDs). Typically, a pcLEDs is formed by combining YAG:Ce phosphor and silicone glue. The YAG:Ce phosphor absorbs the exciting blue light from the chips to stimulate the yellow light, a process that generates white light with desired color temperature [2]. In general, the phosphor-scattered blue light and the phosphor-emitted yellow light have different radiant intensity distributions. As a result, the spatial color becomes non-uniform [3]. Correspondingly, like the conformal phosphor white LEDs which applied in this study, the phenomenon of yellow ring can appear, resulting in disadvantage of vision, see Fig. 1.

Specifically, the blue light during the scattering process is weakened due to the absorption of phosphors, while the energy of converted yellow light is increased after each scattering. Due to the discrepancy of light behavior varying with wavelengths in the phosphor layer, it is necessary and natural to adjust the color performance of pcLEDs. Up to now, there have been studies that propose solutions to improve the color uniformity of LEDs by adding SEP into pcLEDs [4, 5].

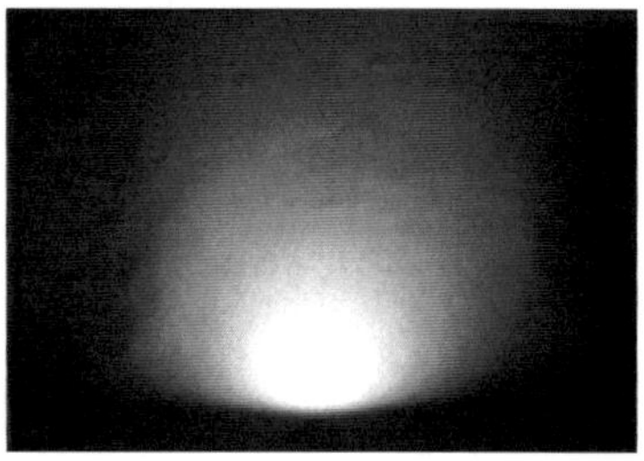

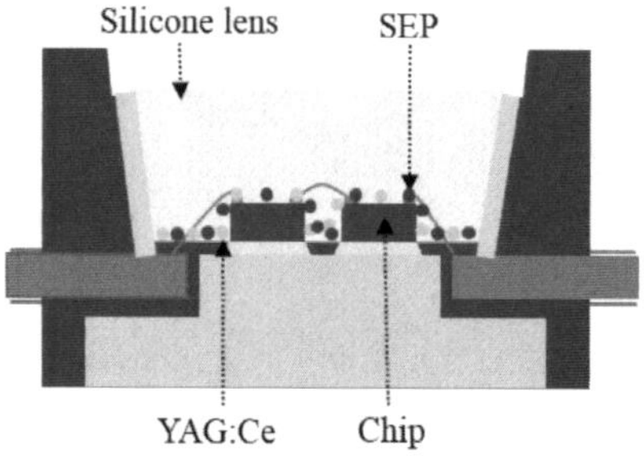

Figure. 1. (top) The yellow-ring phenomenon and (bottom) Illustration of WLEDs

In 2010, the group of Lee mingled titania (TiO_2) and the encapsulation layer as well as the phosphor layer. This study proposed used 0.1% TiO_2 in the encapsulation layer case for obtaining higher color homogeneity (Lee et al. 2010) [6]. In 2013, with the purpose of color homogeneity enhancement, Yang et al. (2013) applied $CaCO_3$ for enhancing the scattering property of WLEDs. The obtained results indicated that the spatial color homogeneity can be increased with 10% $CaCO_3$, significantly [7].

From the work of Anh et al. [1], it has been known that by mixing SiO_2 particles to pcLEDs, the spatial color uniformity can be controlled. Moreover, some evidence has been found to confirm a link between the location of SiO_2 particles in the silicone layer and the color performance. The color temperature of pcLEDs has also been influenced strongly by the size of SiO_2 particles [2].

The above studies have addressed the SEP that can affect the color performance of WLEDs specifically. However, how to select a suitable material among SEP for higher color homogeneity and higher luminous efficiency is still a query. Moreover, the previous studies indicated that the SEP can decrease both the deviation of correlated color temperature (CCT) and the luminous efficiency of the simple single-chip packages having low CCT. In fact, some SEP can enhance the luminous efficiency with a suitable concentration and size.

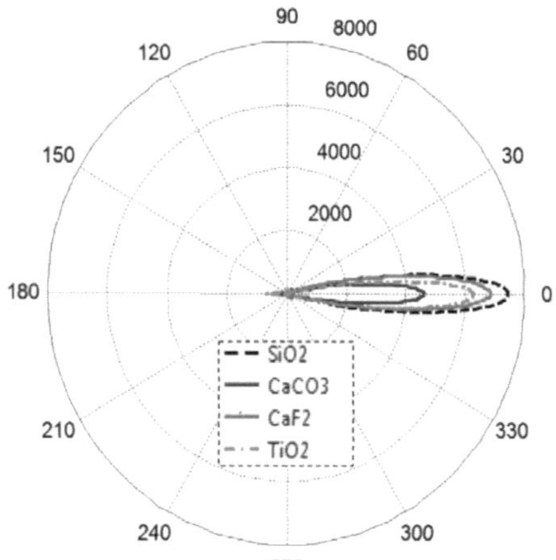

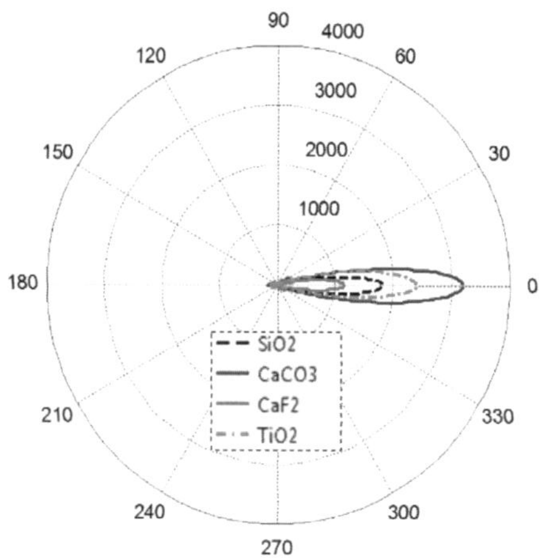

Figure. 2. Angular scattering amplitudes of SEP for 455 nm (top) and 595 nm (bottom)

In this study, we provide an insightful inspection of various scattering enhancement particles including $CaCO_3$, CaF_2, SiO_2, and TiO_2, which are popularly employed for manufacturing higher-quality WLEDs. The target of this work is not only limited on how to select the best SEP but also demonstrated why the SEP can be done by using the Mie theory and the Mote Carlo simulation. The rest of this paper is organized as follow. The basic analysis of scattering inside WLEDs is provided in Section 2. The detailed optical experiments to achieve the goals of this paper as well as some discussion on the simulation results are described in Section 3. Finally, Section 4 concludes the paper.

2. Scattering analysis

The light scattering effect caused by the SEP in the used conformal phosphor WLEDs structure is determined by using Mie-scattering theory [8] with the support of MATLAB as a computation tool. Recall that the scattering amplitude functions $S_1(\theta)$ and $S_2(\theta)$ is determined as following can be calculated by the following equations:

$$S_1 = \sum_{n=1}^{\infty} \frac{2n+1}{n(n+1)} \left[a_n(x,m)\pi_n(\cos\theta) + b_n(x,m)\tau_n(\cos\theta) \right] \tag{1}$$

$$S_2 = \sum_{n=1}^{\infty} \frac{2n+1}{n(n+1)} \left[a_n(x,m)\tau_n(\cos\theta) + b_n(x,m)\pi_n(\cos\theta) \right] \tag{2}$$

$$a_n(x,m) = \frac{\psi_n'(mx)\psi_n(x) - m\psi_n(mx)\psi_n'(x)}{\psi_n'(mx)\xi_n(x) - m\psi_n(mx)\xi_n'(x)} \tag{3}$$

$$b_n(x,m) = \frac{m\psi_n'(mx)\psi_n(x) - \psi_n(mx)\psi_n'(x)}{m\psi_n'(mx)\xi_n(x) - \psi_n(mx)\xi_n'(x)} \tag{4}$$

$$\pi_n(\cos\theta) = \frac{P_n^{(1)}(\cos\theta)}{\sin\theta} \tag{5}$$

$$\tau_n(\cos\theta) = \frac{dP_n^{(1)}(\cos\theta)}{d\theta} \tag{6}$$

where:

- x is the size parameter
- m is the refractive index
- θ is the scattering angle (degree)
- a_n and b_n are in turn the expansion coefficients with even symmetry and odd symmetry
- $\pi_n(\cos\theta)$ and $\tau_n(\cos\theta)$ describe the angular scattering patterns of the spherical harmonics

The angular scattering amplitudes of SEP are computed by MATLAB program, see Fig. 2. The results show that SEP are of great advantage to blue-light scattering. As compensating blue-light enough, it is not only reduce the yellow ring phenomenon, but also increase lumen output. This depend on the selection of SEP, which has approximate angular scattering amplitudes between blue-light (455 nm) and yellow-light (595 nm). Unlike SiO_2 particles, the angular scattering amplitudes of $CaCO_3$ particles present a smallest difference between 455 nm and 595 nm.

3. Simulation and discussion

In this section, the optical properties of WLEDs that employ different SEP is evaluated by using the LightTools 8.1.0 software. The refractive index of the SEP including $CaCO_3$, CaF_2, SiO_2, and TiO_2 are 1.66, 1.44, 1.47, and 2.87, respectively. All SEP are assumed to be spherical with radius of 0.5 μm. The average radius of a phosphor particle is 7.25 μm and its refractive index is equal to 1.83 at all wavelengths in visible band. The refractive index of the silicone glue is 1.5. The diffusional particle density is varied for optimizing the CCT uniformity and the output efficiency.

$$W_{phosphor} + W_{silicone} + W_{SEP} = 100\% \tag{7}$$

$W_{phosphor}$, $W_{silicone}$ and W_{SEP} are the weight percentages of the silicone, phosphor and SEP of the phosphor layer in WLEDs, respectively. If the weight percentage of the SEP is increased, the weight of YAG:Ce phosphor needs to be reduced to maintain the mean CCT value at 8500K. To evaluate the light quality of WLEDs, we determine how much different between CCT values in proportion to angles. This is an important standard for evaluating the performance of solid-state lighting applications. A large deviation in CCT values in proportion to angles will cause the yellow ring phenomenon and generate the non-uniform white color at different angle. The angular CCT deviation is calculated as

$$D\text{-}CCT = CCT_{(Max)} - CCT_{(Min)} \quad (8)$$

where $CCT_{(Max)}$ and $CCT_{(Min)}$ represent the maximal CCT at the zero degree and the minimal CCT occurs at the 70 degree of viewing angle, respectively.

In fact, the reason that causes variation in optical properties of WLEDs is due to the difference among the scattered light of each particle in pcLEDs. The key idea is that this CCT deviation can be reduced significantly as long as the scattered blue light is enhanced sufficiently. As displayed in Fig. 2, the deviation of angular scattering amplitude of $CaCO_3$ particles for 455 nm and 595 nm is smallest in comparison with the other SEP. It means that there is a smallest variation in radiant intensity distributions for the phosphor-scattered blue light and the phosphor-emitted yellow light as using $CaCO_3$.

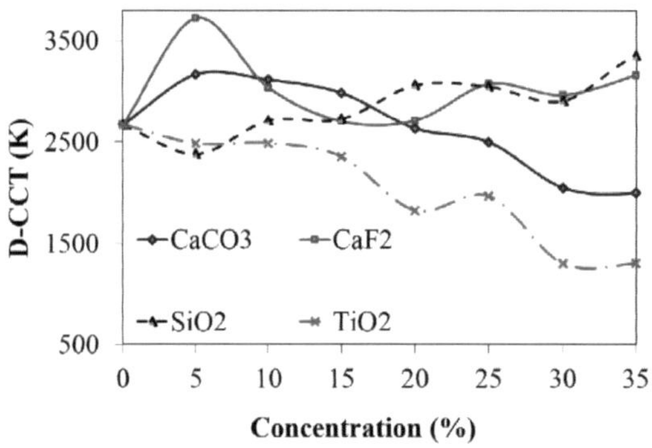

Figure. 3. Comparisons of CCT deviations of WLEDs among SEP

Moreover, like all the applied SEP, the angular scattering amplitude of $CaCO_3$ particles for 455 nm is higher than 595 nm The scattered blue light not only combines with the converted yellow light, but also with the yellow ring for emitting white light, which would alleviate the yellow ring phenomenon in WLEDs. Conversely, the CCT deviation will become larger under condition of lack or redundancy of scattered blue light in WLEDs. Also indicated in Fig. 2, the deviations of angular scattering amplitude for 455 nm and 595 nm can be occurred to 4 times with SiO_2, CaF_2 and TiO_2 particles. Therefore, this is not of much benefit to control the color uniformity and the lumen output.

All discussion above can be illustrated in Fig. 3, in which the CCT deviations for $CaCO_3$ and TiO_2 cases have a tendency to go downward. Specially, the angular CCT deviation can be decreased significantly by using TiO_2. For CaF_2 and SiO_2 cases, it's interesting to observe that the CCT deviations in both cases grow with the concentration of corresponding particles.

4. Conclusion

In this paper, we analyze the effects of various scattering enhancement particles on two optical properties of WLEDs for achieving higher color uniformity. Based on the Mie-scattering theory and the Monte Carlo simulation method, we can verify that the scattered light enhancement achieved by using each type of particles in pcLEDs is quite different from each other. This opens a new door to access some methodology for effectively controlling of optical properties of WLEDs by selecting appropriate scattering enhancement particles with appropriate concentration. We found some interesting results that can serve as a guideline for practical manufacturing of W-LEDs to meet some pre-defined specification on light quality. The CCT deviation would reduce when the concentration of $CaCO_3$ and TiO_2 increases. In particular, the best color uniformity of W-LEDs can be obtained in TiO_2 case.

Acknowledgments
This research is funded by Vietnam National Foundation for Science and Technology Development (NAFOSTED) under grant number 103.03-2015.62.

References

[1] N. D. Q. Anh, M. F. Lai, H. Y. Ma, H. Y. Lee, "Enhancing of correlated color temperature uniformity for multi-chip white-light LEDs by adding SiO2 to phosphor layer," J. Chin. Inst. Eng., vol. 38, pp. 297-303, 2015.
[2] M. F. Lai, N. D. Q. Anh, H. Y. Ma, H. Y. Lee, "Scattering effect of SiO2 particles on correlated color temperature uniformity of multi-chip white light LEDs," J. Chin. Inst. Eng., vol. 39, pp. 1-5, 2016.
[3] Y. H. Won, H. S. Jang, K. W. Cho, Y. S. Song, D. Y. Jeon and H. K. Kwon, "Effect of phosphor geometry on the luminous efficiency of high-power white light-emitting diodes with excellent color rendering property," Opt. Lett. vol. 34, pp. 1-3, 2009.
[4] K. J. Chen, C. C. Lin, H. V. Han, C. Y. Lee, S. H. Chien, K. Y. Wang, S. H Chiu, Z. Y. Tu, J. R. Li, T. M Chen, X. Li, M. H. Shi and H. C. Kuo, "Wide-Range Correlated Color Temperature Light Generation From Resonant Cavity Hybrid Quantum Dot Light-Emitting Diodes," IEEE J. Sel. Top. Quantum Electron, vol. 21, pp. 1-7, 2015.
[5] M. Zhang, X. H. Li, Z. X. Wang, Q. Y. Hu, H. J. Guo, "Synthesis of Y2O3:Eu3+ phosphors by surface diffusion and their photoluminescence properties", Trans. Nonferrous Met. Soc. China, vol. 20, pp. 115-118, 2010.
[6] K.-C. Lee, S.-M. Kim and J.-H. Moon. 2010. "The effects of titania diffuser on angular color homogeneity in the phosphor conformal coated white LEDs," Proc. SPIE 7784, Tenth International Conference on Solid State Lighting, Aug. 2010: pp. 7784101-7784106, doi: 10.1117/12.861108.
[7] L. Yang, Z.-C. Lv, Y. Jiaojiao and S. Liu, "Effects of melamine formaldehyde resin and CaCO3 diffuser-loaded encapsulation on correlated color temperature uniformity of phosphor-converted LEDs," Appl. Opt. vol. 52, pp. 5539-5544, 2013.
[8] S. Liu and X. B. Luo, LED Packaging for Lighting Applications: Design, Manufacturing and Testing, John Wiley & Sons, 2011, pp. 67-144.

Plane strain bending under tension of a functionally graded sheet at large strains as an ideal flow process

Sergei Alexandrov[1,2] and **Alexander Pirumov**[3]

Division of Computational Mathematics and Engineering, Institute for Computational Science, Ton Duc Thang University, Ho Chi Minh City, Vietnam
Faculty of Civil Engineering, Ton Duc Thang University, Ho Chi Minh City, Vietnam
Department of Elektrotechnics and Mechanics, Moscow Technological University, Moscow, Russia

sergeialexandrov@tdt.edu.vn

Abstract. Ideal plastic deformations (ideal flows) have been defined as solenoidal smooth deformations in which an eigenvector field associated everywhere with the greatest (major) principal rate of deformation is fixed in the material. In the case of plane strain deformation of rigid perfectly plastic material obeying an arbitrary isotropic yield criterion and its associated flow rule, it is always possible to find an equilibrium stress field which is compatible with an ideal deformation. It is shown in the present paper that an ideal deformation is possible for functionally graded sheets in the process of plane strain bending under tension. In contrast to the general process, the tensile force and bending moment cannot be prescribed arbitrary but should be found from the solution.

1. Introduction

Ideal plastic flows are those for which all material elements follow minimum work paths. The ideal flow theory has long been associated with the Tresca yield criterion [1, 2]. Recently, it has been demonstrated that ideal flow solutions exist in anisotropic plasticity [3]. In particular, the theory of anisotropic plasticity proposed in [4] has been adopted. A comprehensive overview of the ideal flow theory and corresponding solutions has been provided in [5]. This theory has been widely used as the basis for inverse methods for the preliminary design of bulk metal forming processes driven by minimum plastic work (see, for example, [6]). The process of bending under tension is widely used in metal forming technologies. Using a number of assumptions concerning the through-thickness distribution of strains analytic plane-strain solutions at large strains have been proposed in [7, 8]. Ignoring the transverse stress a solution has been obtained in [9]. An efficient semi-analytic approach to analysis of plane strain bending under tension has been proposed in [10]. This approach is an extension of the corresponding approach for the process of pure bending [11]. In the present paper, the approach developed in [10] is used in conjunction with the ideal flow theory to design the process of plane strain bending under tension of functionally graded sheets. Other ideal flow solutions for this process have been proposed in [12, 13].

2. Kinematics

For incompressible materials an approach to analysis of plane strain pure bending at large strains has been proposed in [11]. This approach has been extended to bending under tension in [10]. For isotropic material models kinematics of the process of bending under tension is independent of constitutive equations others than the equation of incompressibility. For completeness, the basic equations derived in [10] are presented in this section. The process of bending under tension is described by the following mapping between Eulerian Cartesian coordinates (x, y) and Lagrangian coordinates (ζ, η):

$$\frac{x}{H}=\sqrt{\frac{\zeta}{a}+\frac{s}{a^2}}\cos(2a\eta)-\frac{\sqrt{s}}{a},\quad \frac{y}{H}=\sqrt{\frac{\zeta}{a}+\frac{s}{a^2}}\sin(2a\eta) \tag{1}$$

where H is the initial thickness of the sheet, s is an arbitrary function of a, a is a function of the time, t, and $a=0$ at $t=0$. The function $s(a)$ should satisfy the condition

$$s=\frac{1}{4} \tag{2}$$

for $a=0$. Using (1) and (2) it is possible to verify by inspection that $x\to\zeta H$ and $y\to\eta H$ as $a\to 0$. The initial shape of the sheet is a rectangular. With no loss of generality it is possible to assume that the sides of this rectangular are given by the equations $x=-H$ (or $\zeta=-1$), $x=0$ (or $\zeta=0$) and $y=\pm L$ (or $\eta=\pm L/H$). It is convenient to introduce a moving plane polar coordinate system (r,θ) with its origin O_1 at $x=-H\sqrt{s}/a$ and $y=0$. Then, it is possible to find from (1) that

$$\frac{r}{H}=\frac{\sqrt{\zeta a+s}}{a},\quad \theta=2a\eta \tag{3}$$

Since (ζ,η) is a Lagrangian coordinate system, it follows from this equation that the shape of the sheet after any amount of deformation is a circular sector. Its concave, $r=r_1$, and convex, $r=r_2$, boundaries are determined from (3) at $\zeta=-1$ and $\zeta=0$, respectively. As a result,

$$\frac{r_1}{H}=\frac{\sqrt{s-a}}{a},\quad \frac{r_2}{H}=\frac{\sqrt{s}}{a},\quad \frac{h}{H}=\frac{\sqrt{s}-\sqrt{s-a}}{a}. \tag{4}$$

Here h is the current thickness of the sheet. The straight boundaries of this sector are given by

$$\theta=\pm\frac{2aL}{H}. \tag{5}$$

The coordinate curves of the Lagrangian coordinate system are principal strain rate trajectories. The principal strain rates can be found from (1) as

$$\xi_\zeta=-\frac{(\zeta+ds/da)}{2(\zeta a+s)}\frac{da}{dt},\quad \xi_\eta=\frac{(\zeta+ds/da)}{2(\zeta a+s)}\frac{da}{dt} \tag{6}$$

It is evident from this equation that the equation of incomressibility, $\xi_\zeta+\xi_\eta=0$, is satisfied. Since $\xi_\zeta=\xi_\eta=0$ at the neutral line, the equation for this line follows from (6) as

$$\zeta=\zeta_n=-\frac{ds}{da}. \tag{7}$$

It will be shown in subsequent sections that this description of kinematics is compatible with stress equations for the model chosen.

3. Material model

The material is supposed to be rigid plastic. In particular, the elastic portion of the strain tensor is neglected. It is assumed that the principal axes of the strain rate and stress tensors coincide. Then, the coordinate curves of the Lagrangian coordinate system are principal stress trajectories. The plane strain yield criterion at the initial instant is

$$|\tau_x| = |\tau_y| = \frac{\sigma_0 g(x)}{\sqrt{3}} \tag{8}$$

where τ_x and τ_y are the deviatoric stresses in the (x, y) coordinate system, $\sigma_0 g(x)$ is the yield stress in uniaxial tension, σ_0 is a constant with dimension of stress and $g(x)$ is a given non-dimensional function of its argument. It is assumed that plastic deformation does not change material properties at material particles. Therefore, using (8) the yield criterion after any amount of deformation can be written as

$$|\tau_\zeta| = |\tau_\eta| = \frac{\sigma_0 g(H\zeta)}{\sqrt{3}} \tag{9}$$

where τ_ζ and τ_η are the deviatoric stresses in the Lagrangian coordinate system. Moreover, it is seen from (3) that

$$\tau_\zeta = \tau_r, \quad \tau_\eta = \tau_\theta \tag{10}$$

where τ_r and τ_θ are the deviatoric stresses in the polar coordinate system. The associated flow rule is automatically satisfied because the mapping (1) satisfies the equation of incompressibility and, by assumption, the principal axes of the strain rate and stress tensors coincide. Thus the only constitutive equation to be satisfied is the yield criterion.

4. General stress solution

The sheet is loaded by bending moment M and force F per unit length. A consequence of the equilibrium equations is that some pressure P should be applied to the concave surface of the sheet in the process of deformation. Assuming that the pressure is uniformly distributed over the surface it is possible to find that $F = r_1 P$. Using equation (4) this relation can be transformed to

$$\frac{P}{\sigma_0} = \frac{fa}{\sqrt{s-a}}, \quad f = \frac{F}{\sigma_0 H}. \tag{11}$$

Let σ_ζ and σ_η be the normal stresses in the Lagrangian coordinates. It is evident that these stresses are the principal stresses. In the case under consideration $\tau_\zeta = -\tau_\eta$. Substituting this equation into the identity $\sigma_\zeta - \tau_\zeta = \sigma_\eta - \tau_\eta$ leads to

$$\sigma_\eta = \sigma_\zeta - 2\tau_\zeta. \tag{12}$$

The bending moment, M, and its dimensionless representation, m, are

$$M = \int_{r_1}^{r_2} \left(\sigma_\eta - \frac{F}{h}\right) r dr, \qquad m = \frac{2\sqrt{3}M}{\sigma_0 H^2}. \tag{13}$$

Note that $m=1$ in the process of pure bending of a homogeneous rigid perfectly plastic sheet if σ_0 is the yield stress in uniaxial tension [14]. Replacing in (13) integration with respect to r with integration with respect to ζ by means of (3) results in

$$m=\frac{\sqrt{3}}{a}\int_{-1}^{0}\left(\frac{\sigma_\eta}{\sigma_0}-f\frac{H}{h}\right)d\zeta. \tag{14}$$

Taking into account (10) the only non-trivial equilibrium equation in the polar coordinates can be written as

$$\frac{\partial\sigma_\zeta}{\partial r}+\frac{\tau_\zeta-\tau_\eta}{r}=0.$$

Replacing in this equation the derivative with respect to r with the derivative with respect to ζ by means of (3) leads to

$$\frac{\partial\sigma_\zeta}{\partial\zeta}+\frac{a\left(\tau_\zeta-\tau_\eta\right)}{2\left(\zeta a+s\right)}=0. \tag{15}$$

The inequalities $\tau_\zeta>0$ and $\tau_\eta<0$ are valid in the region $-1\le\zeta<\zeta_n$. Then, substituting (9) into (15) gives

$$\frac{\partial\sigma_\zeta}{\sigma_0\partial\zeta}+\frac{ag\left(H\zeta\right)}{\sqrt{3}\left(\zeta a+s\right)}=0. \tag{16}$$

The boundary condition to this equation is $\sigma_\zeta=-P$ for $\zeta=-1$ where P is given in (11). The solution of equation (16) satisfying this boundary condition is

$$\frac{\sigma_\zeta}{\sigma_0}=-\frac{fa}{\sqrt{s-a}}-\frac{a}{\sqrt{3}}\int_{-1}^{\zeta}\frac{g\left(H\chi\right)}{\left(\chi a+s\right)}d\chi. \tag{17}$$

Here χ is a dummy variable of integration. Let σ_n be the value of σ_ζ at $\zeta=\zeta_n$. Then, it follows from (17) that

$$\frac{\sigma_n}{\sigma_0}=-\frac{fa}{\sqrt{s-a}}-\frac{a}{\sqrt{3}}\int_{-1}^{\zeta_n}\frac{g\left(H\chi\right)}{\left(\chi a+s\right)}d\chi. \tag{18}$$

The inequalities $\tau_\zeta<0$ and $\tau_\eta>0$ are valid in the region $\zeta_n<\zeta\le 0$. Then, substituting (9) into (15) gives

$$\frac{\partial\sigma_\zeta}{\sigma_0\partial\zeta}-\frac{ag\left(H\zeta\right)}{\sqrt{3}\left(\zeta a+s\right)}=0. \tag{19}$$

The boundary condition to this equation is $\sigma_\zeta=0$ for $\zeta=0$. The solution of equation (19) satisfying this boundary condition is

$$\frac{\sigma_\zeta}{\sigma_0}=\frac{a}{\sqrt{3}}\int_{0}^{\zeta}\frac{g\left(H\chi\right)}{\left(\chi a+s\right)}d\chi. \tag{20}$$

It follows from this equation that

$$\frac{\sigma_n}{\sigma_0} = \frac{a}{\sqrt{3}} \int_0^{\zeta_n} \frac{g(H\chi)}{(\chi a + s)} d\chi. \tag{21}$$

Equations (18) and (21) combine to give

$$-\frac{\sqrt{3} f}{\sqrt{s-a}} - \int_{-1}^{\zeta_n} \frac{g(H\chi)}{(\chi a + s)} d\chi = \int_0^{\zeta_n} \frac{g(H\chi)}{(\chi a + s)} d\chi. \tag{22}$$

In this equation, ζ_n can be replaced with $-ds/da$ by means of (7). If f is prescribed then (22) is an ordinary differential equation for s as a function of a. This equation should satisfy the boundary condition (2). However, in the case of ideal deformation f should be found from the solution.

5. Ideal deformation

It follows from the conditions imposed on ideal deformation (see, for example, [5]) that ζ_n is independent of a. At the initial instant, the distribution of the normal stress along the surface $y = L$ is

$$\begin{aligned} \sigma_\eta &= \sigma_0 g(H\zeta) \quad \text{in the range } \zeta_n < \zeta \le 0 \\ \sigma_\eta &= -\sigma_0 g(H\zeta) \quad \text{in the range } -1 \le \zeta < \zeta_n. \end{aligned} \tag{23}$$

The initial value of F should be prescribed, $F = F_0$. The corresponding value of f is denoted by f_0. Then, it follows from (11) and (23) that

$$\int_{\zeta_n}^{0} g(H\zeta) d\zeta - \int_{-1}^{\zeta_n} g(H\zeta) d\zeta = f_0. \tag{24}$$

This equation determines ζ_n at the initial instant and, therefore, throughout the process of deformation. Substituting this value of ζ_n into (22) gives the variation of f with a that ensures ideal deformation. The variation of m with a is then found from (14) in which the ratio σ_η/σ_0 should be eliminated by means of (7), (12), (17) and (20).

6. Conclusions

A new ideal flow solution for the process of bending under tension of functionally graded sheets has been presented. The through – thickness distribution of the yield stress is characterized by the function $g(x)$ involved in (8) and is quite arbitrary. A review of material property distributions in functionally graded sheets used in the mechanics of functionally graded material structures has been provided in [15]. Any of these distributions can be combined with the new solution. The solution is practically analytic. A numerical technique is only necessary to evaluate ordinary integrals in (14), (17), (20) and (24).

Acknowledgments
The second author was partly supported by Moscow Technological University [16].

References

[1] R. Hill, "Ideal Forming Operations for Perfectly Plastic Solids," J. Mech. Psys. Solids, vol. 15, pp. 223–227, 1967.
[2] O. Richmond and S. Alexandrov, "The Theory of General and Ideal Plastic Deformations of Tresca Solids," Acta Mech., vol. 158, pp. 33–42, 2002.
[3] S. Alexandrov, Y. Mustafa, and E. Lyamina, "Steady Planar Ideal Flow of Anisotropic Materials," Meccanica, vol. 51, pp. 2235–2241, 2016.

[4] I.F. Collins and S.A. Meguid, "On the Influence of Hardening and Anisotropy on the Plane-Strain Compression of Thin Metal Strip," Trans. ASME J. Appl. Mech., vol. 44, pp. 271–278, 1977.

[5] K. Chung and S. Alexandrov, "Ideal Flow in Plasticity," Appl. Mech. Rev., vol. 60, pp. 316–335, 2007.

[6] O. Richmond and H.L. Morrison, "Streamlined Wire Drawing Dies of Minimum Length," J. Mech. Phys. Solids, vol. 15, pp. 195–203, 1967.

[7] M. Yoshida, F. Yoshida, H. Konishi and K. Fukumoto, "Fracture Limits of Sheet Metals under Stretch Bending," Int. J. Mech. Sci., vol. 47, pp. 1885–1896, 2005.

[8] M.H. Parsa and S.N.A. Ahkami, "Bending of Work Hardening Sheet Metals Subjected to Tension," Int. J. Mater. Form., suppl. 1, pp. 173–176, 2008.

[9] M.A. Guler, F. Ozer, M. Yenice, and M. Kaya, "Springback Prediction of DP600 Steels for Various Material Models," Steel Res. Int., vol. 81, pp. 801–804, 2010.

[10] S. Alexandrov, K. Manabe, and T. Furushima, "A General Analytic Solution for Plane Strain Bending under Tension for Strain-Hardening Material at Large Strains," Arch. Appl. Mech., vol. 81, pp. 1935–1952, 2011.

[11] S. Alexandrov, J.-H. Kim, K. Chung, and T.-J. Kang, "An alternative Approach to Analysis of Plane-Strain Pure Bending at Large Strains," J. Strain Anal. Eng. Des., vol. 41, pp. 397–410, 2006.

[12] S. Alexandrov, W. Lee, and K. Chung, "Design of Plane Strain Bending Based on Ideal Flow Theory," Proc. 8th Int. Conf. NUMIFORM2004, American Institute of Physics, pp. 1664-1669, 2004.

[13] E. Lyamina and S. Alexandrov, "An Elastic Plastic Non-Steady Ideal Flow Solution and Its Application to Bending under Tension," J. Technol. Plast., vol. 34, pp. 39–47, 2009.

[14] R. Hill, The Mathematical Theory of Plasticity. Oxford: Oxford University Press, 1950.

[15] I. Elishakoff, D. Pentaras, and C. Gentilini, Mechanics of Functionally Graded Material Structures. Singapore: World Scientific, 2016.

[16] V.L. Pankov, "Effectiveness of Incentive Mechanism and the Potential Level Meeting the Needs of an Employee," Herald of MSTU MIREA, vol. 1, № 4 (9), pp. 288, 2015.

IOP Conf. Series: Journal of Physics: Conf. Series **885** (2017) 012024 doi:10.1088/1742-6596/885/1/012024

Numerical analysis of heat transfer in unsteady nanofluids in a small pipe with pulse pressure

Cheol Park[1], KiTae Yu[2], HeeGeun Song[3], Sedong Kim[4], Hyomin Jeong

[1]Department of Energy and Mechanical Engineering, Gyeongsang National University Cheondaegukchi-Gil 38, Tongyeong, Gyeongnam, 650-160, South Korea;zzong10047@gmail.com
[2]belljidanto@gmail.com

[3]gmlrms1016@gmail.com

[4]opoq123@naver.com

hmjeong@gnu.ac.kr

Abstract. In the present paper, developing turbulence forced convection flows were numerically investigated by using water-Al2O3 nano-fluid through a circular compact pipe which has 4.5mm diameter. Each model has a steady state and uniform heat flux(UHF) at the wall. The whole numerical experiments were processed under the RPM 100 to 500 and the nano-fluid models were made by the Alumina volume fraction. A single-phase fluid models were defined through nano-fluid physical and thermal properties calculations, Two-phase models(mixture granular model) were processed in 100nm diameter. The results show that comparison of nusselt number and heat transfer rate are improved as the Al2O3 volume fraction increased. All of the numerical flow simulations are processed by the FLUENT. The results show an increase from volume fraction concentration and an increase in heat transfer coefficient with increasing RPM.

1. Introduction

Heat transfer fluids play an important role in many industries, including power plants, production processes, transportation and electronics, and the heat transfer characteristics of thermal fluids play an important role in the development of energy efficient heat transfer devices. Past improvements have been based on structural changes and the addition of heat transfer areas, vibrations on the surface of the heat, injection or inhalation of fluids and the application of electrons or magnetic fields. These enhancement techniques rarely meet the increase in heat transfer coefficient and, in this connection, there is a strong demand for the development of advanced heat transfer fluids with higher heat transfer properties. Therefore, a number of recent studies on nanofluids for experimental convective heat transfer through circular tubes have been conducted by senior scholars. However, nanofluid research is still in the hot debate and investigation. In addition, research on the pulse pressure has been actively carried out all over the world, and many research results have been published especially in the field of noise vibration. Non-steady pulsating flow phenomena will be generated by internal smoke and reciprocating compressors. The pulsating pressure in the internal combustion engine will be the

IOP Conf. Series: Journal of Physics: Conf. Series **885** (2017) 012024 doi:10.1088/1742-6596/885/1/012024

silencer part. The function of this silencer is to reduce the pulsation pressure inside the silencer as well as to reduce the back pressure.

Maxwell [1,2] has demonstrated that increasing the volume fraction of solid particles can increase the thermal conductivity of the mixture. The particle diameter currently available for particle manufacturing technology development is smaller than 100 nm, called nano-size. This effort allows the particle mixing fluid to increase the thermal conductivity of the heat transfer fluid by increasing the critical thermal conductivity. In addition, solid nanoparticle colloids are very stable and do not exhibit significant sedimentation under static conditions after weeks or months. In addition, solid colloids float easily when you reduce particle size. Therefore, nanofluidic technology is expected to be suitable for heat transfer fluids. Mohamad and viskanta [3] studied the abnormal natural convection for a low prandtl number of fluids in the cavity, and patterson and Imberger [4] conducted a study on the abnormal natural circulation in a square cavity. Hellims and Churchill [5] numerically studied the natural convection in an unsteady state in a closed vessel. Therefore, this study focuses on the nanofluid only for pulsating pressure in a small tube with a pulse pressure. The purpose of this study is to verify the numerical analysis technique in terms of the fluidity of the nanofluids in a small tube with a pulsating pressure through numerical analysis. The purpose is to secure the data.

2. Mathematical modelling

Geometry and boundary coundition

Fig. 1a shows the actual model of geometry and Fig. 1b is the part of the Two-dimensional Mesh. One of the main purpose of this simulation is trying to understand the near wall behavior of nano-particles, velocity distribution perpendicular to the pipe surface and drag coefficient at the near of pipe wall, so the relatively concentrated mesh in the near wall area is needed. As a result, the mesh has a non-uniform quadratic mesh. The bottom horizontal line is an axis which makes Three-dimensional pipe model at the axisymmetric environment. The upper horizontal line is a wall with Uniform heat flux. The tube consisted with a diameter (D) of 4.5 mm and a length (L) of 1.17 m. Nanofluids is composed of Al_2O_3 particles and water have the physical properties such as Table.1[6]. The total flow rate for each fluid model is input at a constant temperature of 293K and the axial velocity is defined as pressure change over time at 100,300,500RPM and the wall surface has a Heat flux of 5000 (w / m2)

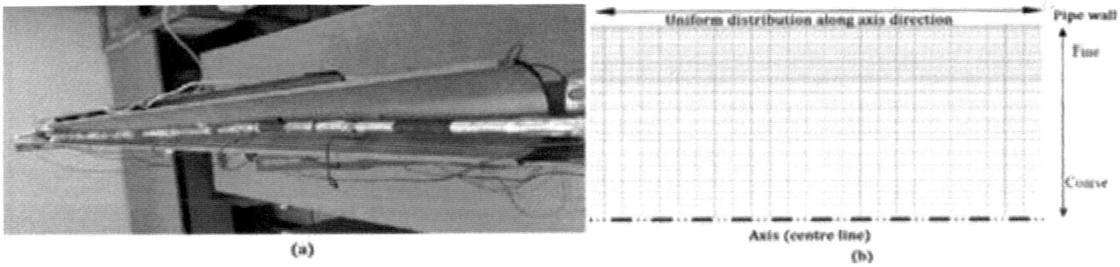

Figure 1. ab Part of the mesh of Computational Cell. 80 x 8000 with 0.96 ratio Governing equation

The single-phase model is investigated for the comparison of mixture model as prescribed. Following equations are representing the mathematical formulation of the single-phase model and of the continuous phase of the mixture model. The exact k-ε equations contain many unknown and unmeasurable term [6-8]

Energy equation:

$$div(\rho\vec{V}CpT) = div(kgradT) + S_e \tag{1}$$

turbulent kinetic energy k

$$\frac{\partial(\rho k)}{\partial t}+\frac{\partial(\rho k u_i)}{\partial x_i}=\frac{\partial}{\partial x_j}\left[\frac{\mu_t}{\sigma_k}\frac{\partial k}{\partial x_j}\right]+2\mu_t E_{ij}E_{ij}-\rho\epsilon \tag{2}$$

dissipation ε

$$\frac{\partial(\rho\epsilon)}{\partial t}+\frac{\partial(\rho\epsilon u_i)}{\partial x_i}=\frac{\partial}{\partial x_j}\left[\frac{\mu_t}{\sigma_\epsilon}\frac{\partial \epsilon}{\partial x_j}\right]+C_{1\epsilon}2\mu_t E_{ij}E_{ij}-C_{2\varepsilon}\rho\frac{\epsilon^2}{k} \tag{3}$$

And k- ε equation of represents velocity component in corresponding direction (u_i), represents component of rate of deformation (E_{ij}) , represents eddy viscosity (μ_t) (Rate of change of k or ε +Transport of k or ε by convection = Transport of k or ε by diffusion + Rate of production of k or ε – Rate of destruction of k or ε)
Conservation of mass :

$$\frac{\delta\rho_m}{\delta t}+\nabla\cdot(\rho_m u_m)=0 \tag{4}$$

Where u_m is the mass-averaged velocity:

$$u_m=\frac{\sum_{k=1}^{n}\alpha_k\rho_k u_k}{\rho_m} \tag{5}$$

and ρ_m is the mixture density with α_k volume fraction of phase *k*:

$$\rho_m=\sum_{k=1}^{n}\alpha_k\rho_k \tag{6}$$

Momentum equation for single-phase:

$$div(\rho\vec{V}\vec{V})=-grad\,P+\nabla\cdot(\mu\nabla\vec{V})+S_m \tag{7}$$

Momentum equation for multi-phase:

$$\nabla\cdot\sum_{k=1}^{n}\alpha_k\rho_k u_k u_k=\nabla\cdot(\rho_m u_m u_m)+\nabla\cdot\sum_{k=1}^{n}\alpha_k\rho_k u_{Mk}u_{Mk} \tag{8}$$

The compression and viscous lose by heat are assumed to be excluded in the energy equation. The source/sink terms S_m and S_e represent the unitive effects of energy and momentum exchange with base fluid. These are zero value in the single-phase model. Momentum equation of multi-phase are affected by volume fraction (α), velocity of phase(u_k), diffusion velocity (u_{Mk}).
The determinations of particle motion are particle mass(m_p), cross-sectional area of the particle(A_p), drag coefficient(C_D), slip velocity(u_{cp}).
The drag force in this study is following Clift et al equation (7) for single rigid spherical particle in a fluid:

$$F_D=-\frac{1}{2}A_p\rho_c C_D|u_{Cp}|u_{Cp}-\frac{1}{2}V_p\rho_c\frac{du_{Cp}}{dt}-6\gamma_p^2\sqrt{\pi\rho_c\mu_c}\int_0^t\frac{\frac{du_{Cp}}{ds}}{\sqrt{t-s}}ds \tag{9}$$

Numerical approach
The computational fluid dynamic code FLUENT 16.1 is employed to solve the present problem. The governing Eqs. (1)-(3) are solved by control volume approach. The algebraic "discretize equations," resulting from spatial inter-fraction process, are sequentially solved throughout the physical domain considered. ANSYS Fluent 16.1 solves the systems resulting from discretization schemes using a

numerical method. For the convective and diffusive terms, a second order upwind method was used. Pressure and velocity were coupled using Semi Implicit Method for Pressure Linked Equations(SIMPLE) in single-phase. To calculate the pulse pressure, Figure 2. the PISO algorithm was used to calculate the anomalous state as a complete negative solution. The continuity equation and the momentum equation are used to calculate the pressure inside the small tube. The turbulence model is based on the standard model k -ε model which has already proved its engineering validity. The convergence judgment of the residuals was regarded as convergence when the residual value reached 10 or less.

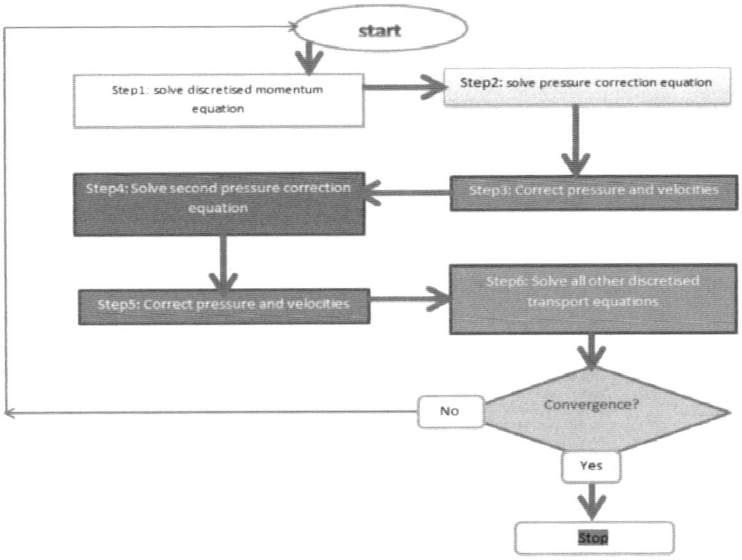

Figure 2. Folw chart of PISO algorithm

As already seen for the SIMPLE algorithm, the steps 4 and 5 can be repeated for a prescribed number of times to correct for non-orthogonality.

3. Results

In this work a number of numerical simulations have been perfomed to study forced convective heat transfer of Al2O3-water in a circular tube under turbulence flow and results were conducted to employing the single phase and mixture model for φ=1% to 4%, 100to500RPM as q=5000 W/m^2 with a constant state. In all cases the size of the spherical particles is 100 nm. Thermal entrance length depends on Prandtl number.[9] So when concentration increases Pr number also increases and consequently, thermal entrance length becomes greater.

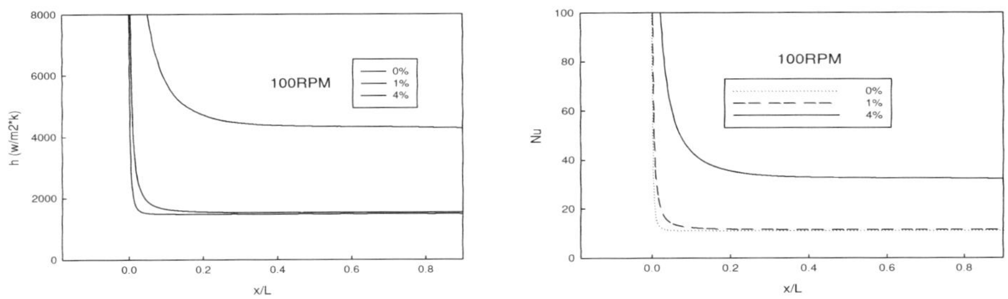

(b)

Figure 3.(a) 100RPM, Heat transfer coefficient for Al2O3-water, (b)100RPM, Nusselt Number

IOP Conf. Series: Journal of Physics: Conf. Series **885** (2017) 012024 doi:10.1088/1742-6596/885/1/012024

As shown in Fig. 3 (a), the difference between 0% and 1% seems to be insignificant in the range of 0.1x / Lmm, but the heat transfer rate of 4% Al2O3 is 2 times larger than 0,1% Respectively. Likewise, in figure (b)

The transmission rate of 0 and 1% Al2O3 is less than 10, but 4% nanofluids can be seen to have more than twice the Nusselt number.

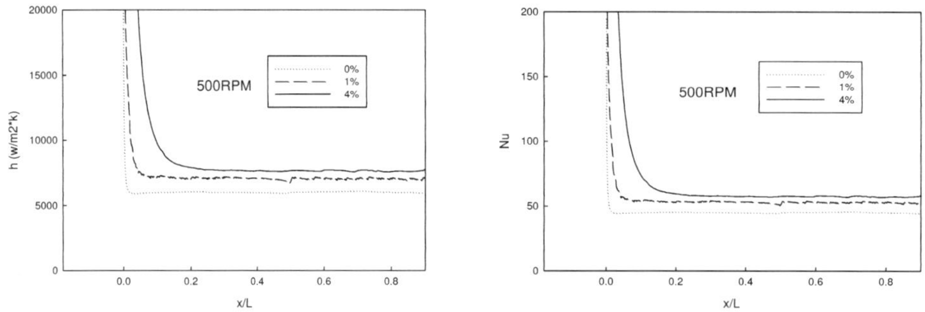

(a) (b)

Figure 4. (a) 500RPM, Heat transfer coefficient for Al2O3-water, (b)500RPM, Nusselt Number

In Fig. 4 (a), unlike 100RPM, the phenomena appearing more than twice as much as 0%, 1% and 4% are not shown in the figure. This phenomenon can be grasped by increasing the internal pressure at 500 RPM. Also, in Fig. (B), the Nusselt number seems to be closer to 1% Al2O3 than 4%.

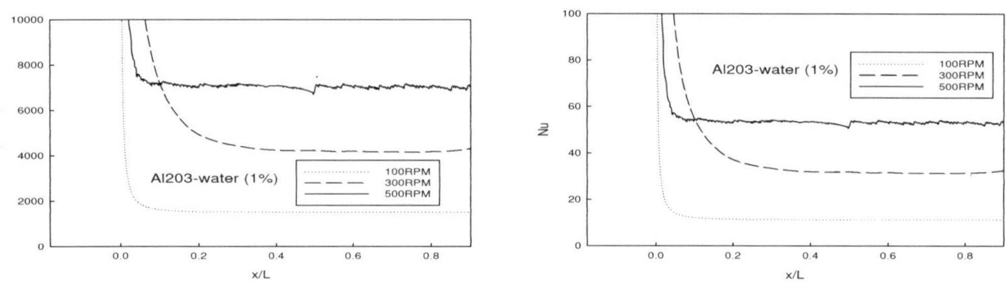

(a) (b)

Figure 5. (a) Al2O3-Water, Heat transfer coefficient for 100 to 500RPM, (b) Al2O3-Water, Nusselt Number for 100 to 500RPM

Figures 3 and 4 show the difference in heat transfer coefficient and number of nuns in the nanofluid for each RPM. 5 (a) and (b), however, the heat transfer coefficient and the number of nuclei for RPM changes of the same 1% Al2O3 can be confirmed. The heat transfer rate and the number of nunts are significantly lower than those of 300 and 500 RPM . However, at 300 RPM, it can be confirmed that the heat transfer coefficient and the nunner are higher than 500 RPM in the early stage.

4. Conclusion

Numerical analysis of the heat transfer of the nanofluid in the unsteady state in the small tube into which the pulsating pressure is introduced showed that the effect of the content of Al2O3, which is a nanofluid, was further increased in the low 100 RPM. In addition, it was confirmed that the nanofluids of 0% and 1% compared to 0% and 1% of Al 2 O 3 have a data difference of more than 2 times. In 500 RPM, the difference in content is not significant compared to 100 RPM, but nanofluids with high Al 2 O 3 The heat transfer coefficient and the number of nuns were better. And the increase in the heat transfer coefficient of the nanofluid to the RPM change shows better results as the RPM increases. Therefore, nanofluid with high Al2O3 content and high RPM can improve the thermoelectric coefficient and increase the number of nuts. But the CFD should be progressed with constant and transient numerical investigation parallel. Basically, It would be pace with the experiment especially different model scale study. Also,uniform heat flux sample, wall shear stress, body force researches does not include this paper. This compact size model is not usual model on the nano-fluid CFD study so these factors need to be proved.

Acknowledgments
This research was supported by Basic Science Research Program through the National Research Foundation of Korea(NRF) funded by the Ministry of Education(No. 2015R1D1A1A01058030).

References

[1] J. C. Maxwell, A Treatise on Electrictiy and Magnetism, Oxford University Press, Cambridge, 1881J. Clerk Maxwell, A Treatise on Electricity and Magnetism, 3rd ed., vol. 2. Oxford: Clarendon, 1892, pp.68–73.
[2] Q. Li, Y. Xuan, Heat transfer enhancement of nanofluids, International Journal 21, pp.58-64, 2000
[3] A.A Mohamad, and R.Viskanta: Int. J. Numerical Method in Fluid, 12, 61-81(1991)
[4] Sunghoon kim,HyeMin Son,JaeCheon Lee, The effect of eccentricity between gear and housing in involute gear pump, Journal of the Korean Society of Marine Engineering 37(6) 2013.9,631-637
[5] Yangwoo Mo,YoungTae Yoo,GeeDae Na,JiHwan Kim, A Characteristics of Impedance Propagation by the Unsteady Flow in a Hydraulic Pipeline, Journal of the Korean Society of Manufacturing Technology Engineers 13(6), 2004.12, 48-55
[6] M. Nazififard, M. Nematollahi, K. Jafarpur, K. Y. Suh, Numerical Simulation of Water-Based Alumina Nanofluid in Subchannel Geometry, Science and Technology of Nuclear Installations, Vol 2012, 2012
[7] ANSYS Fluent Theroy Guide, pp. 501, November 2013
[8] V. Bianco, F. Chiacchio, O. Manca, S. Nardini, Numerical investigation of nanofluids forced convection in circular tubes, Applied Thermal Engineering 29, pp 3632-3642, 2009
[9] V. Bianco, F. Chiacchio, O. Manca, and S. Nardini, Numerical investigation on nanofluids turbulent convection heat transfer inside a circular tube, International Journal of Thermal Sciences, vol. 50, no. 3, pp 341-349, 2011
[10] Chungseub Yi ,kyujin shim,wanda ali akbar,hanshik chung,hyomin Jeong, study on the Pressure Variation in a Chamber Caused by Pulsation Pressure, Transaction of Korean Society of Automotive Engineers 15(4),2007.7, 132-138(7pages)A reference